Believe! The Winning Mindset for Project Managers

Professional life is filled with opportunities to learn. One such opportunity is to learn from colleagues. *Believe! The Winning Mindset for Project Managers* is an opportunity to learn from dedicated project managers.

People who are dedicated and believe in their profession achieve amazing results because they enjoy what they do, feel positive about all activities they do, and are confident and happy with the people around them. The book shows how this is true for project professionals. It is a collection of testimonials from project practitioners who believe in project management. These practitioners are believers who defend with passion, persistence, and patience the project management profession. Their own experiences—both good and bad—provide fodder for growth and insights to illuminate ways to view the power of belief in one's profession. The book features:

- Reasons why identifying beliefs are critical to personal and professional development
- Mindset elements that constitute beliefs
- Examples of manifesting healthy beliefs and behaviors
- Tips, techniques, tools, and lessons learned to apply every day on the job

This book is not about theory. It is about practice and is filled with real case studies, exercises, templates, and lessons learned from the trenches. It explains how belief in oneself as a project manager can accelerate personal growth and career advancement. Through harnessing the power of belief, project managers can boost their energy to accomplish more than they previously thought possible. It is also possible to be more successful in achieving project, as well as personal, goals. Belief in others can lead to greater understanding and empathy with all project stakeholders

This book is an entertaining and productive journey into project management that can lead to not only incremental improvements but also quantum shifts in lifestyle and professional achievements.

Believe! The Winning Mindset for Project Managers

Alfonso Bucero and Randall L. Englund

CRC Press
Taylor & Francis Group
AN AUERBACH BOOK

Designed cover image: Shutterstock

First edition published 2027
2385 NW Executive Center Drive, Suite 320, Boca Raton FL 33431

and by CRC Press
4 Park Square, Milton Park, Abingdon, Oxon, OX14 4RN

CRC Press is an imprint of Taylor & Francis Group, LLC

ISBN: 978-1-041-21975-0 (hbk)
ISBN: 978-1-041-22121-0 (pbk)
ISBN: 978-1-003-73073-6 (ebk)

DOI: 10.1201/9781003730736

Typeset in Minion
by Newgen Publishing UK

Dedication

I (Bucero) want to thank my project management colleagues worldwide, for example to Elisabet Duocastella, PMP who has been a great contributor to our profession over the years and who believes in the profession as a practitioner and also as a PMI volunteer helping do social good. She invited me to speak about the "believe subject" at PMI Barcelona Chapter event in 2024, where my presentation helped to reflect upon my professional story as a believer. I need to share with other professionals to inspire them in our profession. I also thank our friend and colleague Jack Duggal who always transmits his passion as a project practitioner, to Jim De Piante who shares with us his experiences and learnings with passion, to David Hillson (well-known as Doctor Risk) who every time we meet we learn something, to Michel Thiry who recognizes my positive attitude in front of other colleagues, and to Olivier Lazar who underlines my positive project attitude and appreciates my passion.

I (Englund) want to add that today is a good day to thank all project managers and worldwide project professionals who contributed their opinions, experiences, and practices to this and other projects, directly or indirectly. I am deeply grateful to Alfonso for his leadership on this project and everlasting friendship and am grateful for all the opportunities to share and learn from others throughout my career—the practice fields created bountiful harvests. Mark Hamilton provides inspiration in his quest for integrated thinking, self-leadership, and immortality.

This book is dedicated to:

- All leaders (project management students, beginners, senior practitioners, and executives) with intention to inspire them about project management value.
- To Taylor & Francis editors who sponsored us on this project, offering their commitment and support from start to finish of this project.
- To our wives (Rose and Marilyn) for their love and respect. They always support and encourage us in all endeavors.

- To the late Dr. J Davidson Frame and his everlasting support and contributions.
- To my (Bucero) grandson, Manuel De Miguel Bucero, only six years old, with my best wishes of being a personal and professional believer.

We encourage you to share with us your opinions, ideas, feedback, and real stories in order to improve future editions of this book.

Contents

Preface

What is holding you back from accomplishing more in your life and career? Are you looking for a key ingredient that is within yourself and you are capable of accessing? More often than not, your beliefs are both the culprit and the magic ingredient.

Professional life is plenty of opportunities to learn. It is amazing the vast range and number of possibilities to achieve goals if we believe in them, if we think seriously that we can accomplish them. People who believe in their profession achieve amazing results because they enjoy what they do, because positivism is ever-present in all activities they do, and because they are confident and happy with the people around them.

Amazing results means amazing efforts from believers. We found many stories to share with you about believe. The first question we need to answer is "do I believe in myself?" Am I aware of the great possibilities around me, my career, my colleagues? One of my [Bucero] preferred sentences is "you could be my teacher". I am always open to learning from anybody, anyplace, under any circumstances. Are you ready to do that?

In this book we collected testimonials from project practitioners who believe in project management. We met them regularly and keep in touch with them because they are, like us, project management believers who defend our profession with passion, persistence, and patience. Our own experiences (both good and bad) provide fodder and insights to illuminate ways to view the power of our beliefs.

This book is not about theory; it is about practice, real case studies, exercises, templates, and lessons learned from the trenches.

Features found in this book:

- Reasons why identifying beliefs are critical to personal and professional development
- Mindset elements that constitute beliefs
- Examples of manifesting healthy beliefs and behaviors

Benefits:

- Less energy required to accomplish more than previously thought possible
- Accelerate personal growth
- Career advancement

Advantages:

- Greater satisfaction in all endeavors
- Strengths that provide competitive advantages
- Consistent performance in achieving desired results

Please join us in what we believe is an entertaining and productive journey that provides not only incremental improvements but also offers quantum shifts in your lifestyle and achievements.

Alfonso Bucero
BUCERO PM Consulting
Madrid, Spain

Randall L. Englund
Englund PM Consultancy
St. George, USA

About the Authors

Alfonso Bucero, PhD, PMP, PMI-RMP, PfMP, PMI Fellow, is now an independent project management consultant and speaker. He was founder, partner, and director of BUCERO PM Consulting in Spain (www.abucero.com). He was managing director at IIL Spain in Madrid, serving as a project management consultant and training executives, explaining the need for and advantages of project sponsorship. His background was as a project manager at Hewlett-Packard Consulting, where he developed and managed the PMO implementation whose purpose was the continuous improvement of project management discipline across the organization. During his 13 years at HP, he managed various customer, infrastructure, development, and change management projects. He spent his last two years at HP selling and implementing the project office and convincing upper managers about the advantages of project sponsorship. Alfonso has a Computer Science Engineering degree, is a PhD recipient, and has 35 years of practical experience managing and consulting on projects worldwide. As a project management believer, he defends passion, persistence and patience to achieve project success.

From managing projects across Europe, Alfonso shares many real case studies and a fun attitude with seminar participants. In addition to co-authoring *Project Sponsorship* and *The Complete Project Manager*, Alfonso authored *The Influential Project Manager: Winning Over Team Members and Stakeholders* (CRS Press, 2015), *Today is a Good Day: Attitudes for Achieving Project Success* (Multi-Media, 2010), *Tomorrow will be Better* (Motivational Press, 2017), *Passion, Persistence and Patience* (BEP, 2019), and *Today is a Great Day* (CRC 2025). He served as an international project management assessor for the International Project Management Association, sponsor and then president of the Project Management Institute's Barcelona and Madrid chapters, and a frequent contributor to international project management conferences and project office workshops. He was contributing editor of the "Crossing Borders" column in the Project Management Institute's *PM Network* magazine. He is the author of the Spanish book *Project Management: A New Vision* and contributed a chapter to Englund, Graham, and Dinsmore's *Creating the Project Office*. PMI awarded Alfonso with the Distinguished Contribution Award in 2010, the PMI Fellow Award in 2011, and the PMI Eric Jenett Excellence Award in 2017.

Randall L. Englund, MBA, BSEE, NPDP, CBM, as executive consultant for the Englund Project Management Consultancy (https://englundpmc.com), provides management and leadership awareness for clients worldwide at a systemic level through his presentations, workshops, seminars, consulting engagements, books, conference papers, blogs, and other writings. His interactive, multimedia presentations focus on mindset and people skills that create more effective working environments. Experiences stem from 22 years at Hewlett-Packard, where he was a senior project manager at a corporate level and then as an independent consultant for 25 years. He also served as a program manager in high-tech new product development, military officer, and as a field installation supervisor for GE.

He co-authored *Creating an Environment for Successful Projects*, *Creating the Project Office*, *Project Sponsorship*, *The Complete Project Manager*, and *The Complete Project Manager's Toolkit*. In addition to updated, revised editions of these books, he also contributed chapters to other business/management books. He received PMI's Distinguished Contribution Award in 2013, citing a great contribution to the profession worldwide, presenting, consulting, coaching, and mentoring project managers, and convincing executives about the added value of project management implementation in organizations. PMI honored Randy with the Project Management Jenett Award of Excellence in 2018 for significant and lasting contribution to professional services. He also developed and taught graduate certificate online courses *Leading and Managing Technical Projects* for Northeastern University

FIGURE 0.1
Alfonso Bucero and Randall Englund.

College of Professional Studies since 2015 and *Management, Leadership, and Team Building in the Project Environment* for UC Irvine Extension in 2014. His ongoing purpose is to change thinking and apply organic concepts that create environments for successful projects, where people feel empowered to do their best work.

The authors bring a practical approach to this book—the point of view of people who have been through it all. Alfonso and Randy are project management practitioners and use passion, persistence, and patience when implementing project management in organizations. Their ongoing commitment is to bring forth lessons from productive movement between reflection and action. Both authors put effort into systematic thinking about project management beliefs, sharing ideas, and interviewing colleagues worldwide. This book was written because of the need from the authors to share their beliefs about project management to inspire new professionals and experienced practitioners as well. We believe people are winners who tell the best stories: our story is *experienced facilitators seek ways to make a difference in lives of colleagues. One approach is discovering organic metaphors that enhance learning. Another approach is fueling Passion, Persistence, and Patience.* We need more and more project management believers, as the world needs this.

1

Introduction

***Your brain** is like a sleeping giant.*

Tony Buzan

INTRODUCTION

I (Bucero) always was interested in "why do I like certain things?" I believe it is a good way to learn from your brain. I love studying about the brain. How it works and how our emotions and thoughts affect each other is fascinating to me.

Do you have thoughts you want to get out of your head? Maybe you say or do things that you want to change? We all do but can we really change? For a long time, scientists have believed that our brains are the way they are and that's just it. They are made that way when we are young and they stay that way. And there's no way to change that. Thankfully, science is constantly improving and sometimes they even figure out that they are completely wrong. Now, scientists know that the brain can change and that we are not set in stone. We can change. We can grow. We can transform our minds. As project management professionals we need to renew our minds. The word transform means to "make a thorough or dramatic change in the form, appearance, or character" (dictionary.com). In this case we are commanded to make a thorough or dramatic change in our mind, our thinking. We can transform our thinking and what we believe. But how do we do that?

DOI: 10.1201/9781003730736-1

What Do We Know About Our Brain?

Along with many other things, our minds contain neurons, but for now, let's just concentrate on them. Certain parts of the brain will light up when doctors scan patients' brains and ask them queries. The electrical transmissions between neurons are what those "lights" are. The ideas are in there. Simply stated, neurons are what make up our thoughts because they are activated when we think. See Figure 1.1.

New synapses are created as we develop and learn. Some religious practitioners think of how God created everything using only his words. When we learn something new, we generate new neurons solely with our ideas. (You could have just made a new synapse at that moment.) What a great idea. Then, through the dendrites and axons of what is known as a synapse, the new neurons link to the pre-existing neurons, or pre-existing thoughts. When we think about or recall something, a neuron experiences a stimulus, which is then transmitted to a connecting neuron, which in turn triggers another thought. Have you ever wondered why your thoughts wander? You think of a plane which makes you think of the sky which makes you wonder why the sky is blue which makes you think of how you argued about the color sky blue or light blue in second grade, then you think of the kid in second grade who picked his nose … and the thoughts continue. All those thoughts happen in a split second because of the electrical signals between neurons. It's amazing what our brains can do.

Additionally, just as our brain produces new neurons, it can also destroy existing ones. These are no longer active neurons or ideas. You may be familiar with *Inside Out*. You ought to view it if you haven't already. It's a fantastic drama about the brain. The plot of that film centers on Riley's internal thoughts and memory-making processes. Some of these recollections endure for years, while others fade into obscurity. In the film, these two men roam

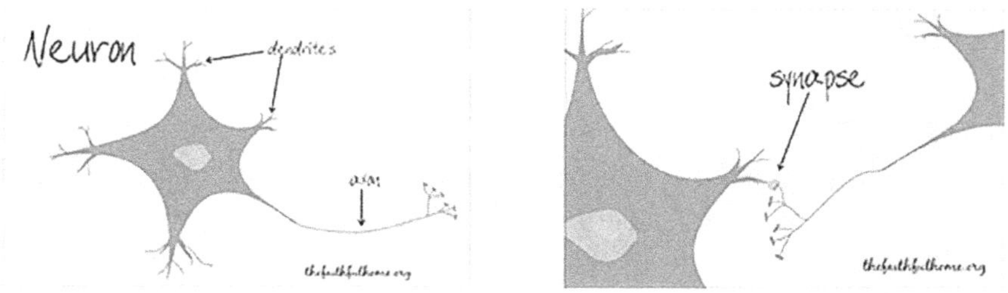

FIGURE 1.1
Neuron and synapse.

Riley's mind, removing the murky recollections with a vacuum. They must go because Riley is no longer using them.

Our minds function very similarly. Our brains contain substances (hormones) that eliminate inactive neurons. If you've ever attempted to recall something but were unable to, that may be the cause. Because you don't use it frequently, the cell is deteriorating. The neurons that we use more frequently are more interconnected with one another, and they also carry more electrical impulses and hormones.

Our Brains Are Awesome and Do Cool Things, But Who Cares?

Consider school. The smell of crayons and construction paper, the instructor, the students, the projects, the woven naptime rugs, chalk squeaking on blackboards. Your neurons are responding when you think of school. However, there is also something else going on. Your hormones in your brain are triggering your feelings. Because of hormones, we literally feel when we ruminate about something.

Consider now visiting a dentist to have a cavity filled. Your tongue is numbed by the needle, keeping your mouth open when someone uses instruments to press and scrape. Your brain would appear to be lit up like a Christmas tree on a brain scan due to the feeling associated with having a cavity filled.

Two distinct ideas, two various emotional reactions (unless you had an awful kindergarten teacher in which case it might be the same emotional response). Neurons in our brains represent our thoughts, and hormones in our brains represent the emotions that travel throughout our bodies. What we believe impacts how we feel, which in turn influences what we say and do. This is why the whole cerebral thing is important. It matters what we believe and how we think.

What would we talk about if we thought on "such things" instead of all the things that bother us or anger us or just the things of this world? How would we act if our thoughts were on what is true, noble, right, pure, lovely, admirable, excellent, and praiseworthy? These are questions that we need to consider because how we live comes directly from what we think which comes from what we believe to be true. How do we get our minds to transform and "think about such things"?

Transforming our minds is not something that will happen by accident. It takes daily, persistent, and patient effort to get up every morning and think about the things that are true and right and pure and lovely. This is not a casual

"Oh, I really need to stop thinking that way" kind of thing. Transformation is deliberate. Yes, we can transform our minds and change to be lovely people. It will definitely take work, but it will be so worth it.

Go Ahead and Start Today

What is one thought you need to get out of your mind? Maybe you know what it is right away or maybe not. Maybe you have too many and don't know where to start. That's okay. Starting with reflection about yourself, your acts, is better anyway. Whether you have a thought in mind to get rid of or not, reflect upon that thought and your mind will show you exactly what you need to work on. And when you know, write it down specifically. Like this … if you say, "I need to work on not being angry," well, that's too general. Think: I need to not get angry with my kids or I need to stop being frustrated with slow drivers. Choose one situation. One thought at a time.

What we believe matters. It matters for our lives but more importantly it matters to our brain. So, let's transform our minds. See Figure 1.2.

This book aims to integrate different pieces to develop and sustain your professional beliefs.

FIGURE 1.2
What we believe.

Outline

Each chapter highlights thoughts and real stories about a particular subject linked to our professional careers as project practitioners. Included at the end of each chapter is a Tool or Checklist to guide your thinking and assess progress. Here is an outline of what is covered:

- **Chapter 1**: In this chapter we explain our insights about why to believe in.
- **Chapter 2**: This chapter establishes the foundation, our vision and the believe concepts and the synonyms we found.
- **Chapter 3**: In this chapter we talk about what is the value of "believe in," their components, and how to measure it.
- **Chapter 4**: It is about how much you believe in the profession, in yourself, in your team, in your organization.
- **Chapter 5**: This chapter describes our vision about developing your project management beliefs, assessing your beliefs, and sharing some of our best practices.
- **Chapter 6**: It is about how to maintain and sustain your PM belief in difficult or challenging situations, by providing real stories and examples.
- **Chapter 7**: This chapter explains a project manager mindset. We share best practices about a mindful project manager.
- **Chapter 8**: This chapter summarizes our findings and best practices explained in this book.

Figure 1.3 depicts a mind map with descriptions of chapter titles. Throughout the whole book, we emphasize the importance of being positive and developing a positive attitude when trying to develop your beliefs.

All the best practices and ideas shared in this book are based on our particular experiences and on interactions with other project professionals worldwide. Only we are responsible for contents in this book. Figure 1.4 shows three elements of our beliefs.

You are probably familiar with the iron triangle, also known as the triple constraints for a project. We want to define my BELIEVE triangle for project success. The sides of our triangle are: Passion, Persistence, and Patience. These are three basic skills to be developed in order to believe in yourself, your project team members, and stakeholders. It takes time and practice, but you can do it because you need to be *excellent* as a project practitioner.

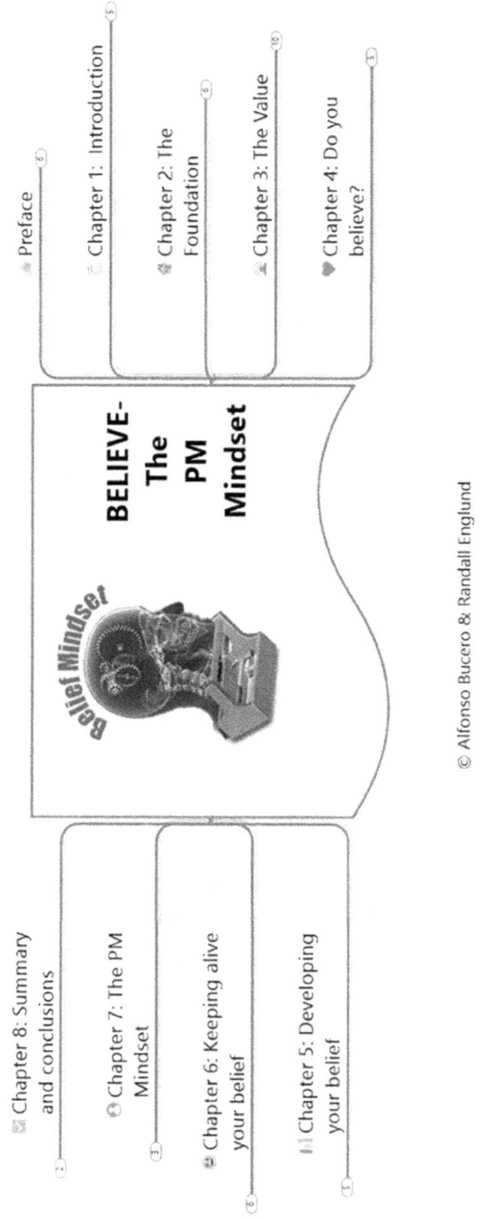

FIGURE 1.3
Believe—the PM mindset.

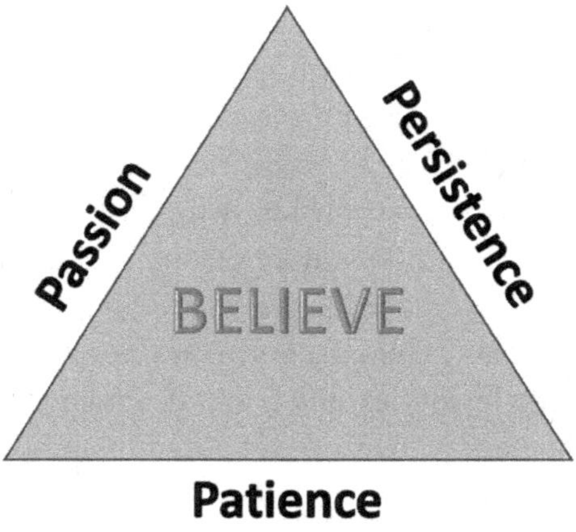

FIGURE 1.4
Three elements of our beliefs.

Did you enjoy your reading so far? We hope so. If your answer is positive, please continue reading Chapter 1. If not, please close the book and try it again tomorrow because we need great professionals like you who are able to give us feedback about your project professional beliefs.

OUR PROJECT MANAGEMENT STORIES

Professional development is about improving yourself through learning, practising, making mistakes, overcoming them, and training. A project professional usually works on their own professional development independently. Human Resource departments may facilitate access to some training or development alternatives, but it is about you spending time and effort exploring your possibilities. Professional development can also refer to keeping abreast of new trends within your field and applying new practices to your current methods.

I (Bucero) am blessed because my professional career, which I consider my path, even when it has not been a path of roses, helped me find many obstacles that allowed me to learn and develop my skills to move forward:

I (Bucero) started my project management career by accident. I was an IT technical specialist and suddenly the company I was working for (SECOINSA) assigned me as a project manager for a key project within a Spanish Bank. I had a fuzzy idea about what a project meant but I found myself trying to lead a project (one year long) with six people in my team, ten years older than me, and most of them with high technical experience. I participated in the planning stage, but I followed the experience of the sales specialist who sold the project (he had been an IT technical specialist in the past). Although I was shocked because of my responsibility for leading that project, I listened to my team and asked them for advice on many occasions. I was 24 years old, and I needed to take the initiative convincing my team to follow me. I ran into problems, but my team supported me and finally this project was a success. It was my first lesson learned about dealing with people as a project manager. I asked them many questions, I asked them how they felt (we were traveling for one year over Spain), I asked them for advice before making a project decision, and I learned. Step by step I built a personal relationship, and they followed me.

After two and a half years I moved to DEC (Digital Equipment Corporation) where I worked in the pre-sales department for one year. During this period, I learned the importance of establishing a strong relationship with salespeople. I learned how to present a customer proposal, how to deal with customer objections, how to get buy-in from the customer, and mainly how to work with the sales representative to win the proposal and move forward with a project. During the two following years, I was assigned as a team leader for software development projects in the database field. In that period, I started dealing with third parties and working well with them. I knew the importance of involving them in the project from the very beginning and counting on them to move forward. I started to use my first project management methodology and developed my project management discipline. Three years later I moved to ICL (International Computers Limited) as a project manager and started to manage projects for several customers in Spain. I gained experience as a project manager with small teams in that organization and started to learn how to deal with executives in projects. I discovered the need to use the language executives understand and the importance of listening well to my customers. A couple of years later I moved to HP (Hewlett Packard) as a project manager, and I was assigned to an important project in the north of Spain. I had to lead a team of 16 people from two different subcontractors over two years. In that project I learned the importance of believing in yourself and being positive most of the time. Despite running into difficulties in that period because my father passed away (lung cancer), I learned how to smile in front of people every day and working together to achieve project objectives. I was outside home for two years, coming back home for one day only and returning to the project site. I learned the importance of establishing personal relationships with my team and my subcontractors. The project was delayed a little bit but was considered a success because we achieved the project scope and cost, and most importantly, the customer used the final product successfully.

I worked for HP almost 14 years as a project manager. I learned, sold, and demonstrated to HP managers the value of project management. I demonstrated through good results the added value of project management, not only by good planning and execution, but by analyzing results and not reinventing the wheel. HP gave me the benefit of the doubt and assigned me as a project manager to implement a PMO (Project Management Office). At HP I strengthened my belief in project management. I dealt with customer executives and internal executives, having the opportunity to learn from them and training them on the importance of executive support. Before leaving HP, I dealt with the most important project of my life, an IT and change management project where I dealt with five third parties and built a strong relationship with customer executives, mainly with the customer sponsor. I demonstrated to my organization the importance of executive involvement and support to achieve project success. I also learned the importance of developing my soft skills as a project manager.

After almost 14 years I left HP, and as a project management believer I was representing IIL (International Institute for Learning) for almost two years, selling and delivering consulting and training services internationally. Then I left IIL and created my own company, BUCERO PM Consulting, that has been operating till now. Project management is my life, I believe in the value of project management, and I am aware of the big difficulties a project manager has in his/her career, but nothing is impossible if you believe in what you are doing.

I enjoy managing projects, coaching, and mentoring project managers and helping professionals and organizations to improve their attitude to managing projects in organizations. I enjoyed writing several books in my career because I strongly believe I was born to learn and share my experiences with others to succeed. This is the main reason I made the decision to write this book. I want to expand the need of being a project management believer. As I said before, I believe in the profession and guess will never be retired because I need to feel alive, too many projects to manage, and too many ideas to convert into reality.

One of my recent projects is my PhD. I started my PhD studies several years ago, finding many obstacles in the path but overcoming them one by one. Now I am successful. I have defended my dissertation and received my PhD.

But what will be my next project? In my talks and presentations, I always say that my life expectancy is 240 years. Now I'm under 70 so I still have many projects to manage and many people to help. It's my way, I always try to be positive and enjoy my life.

My suggestion is please live every day on your project as if it were the last one. Connect with people, listen to them, work with them, forgive their mistakes, and encourage them to learn and improve. You are an excellent professional, meaning by excellent to be able to move the wheel of your efforts ahead and learn more and more every day. You can do it. Go for it!

I (Englund) started out in engineering because I was good at math and science. My college education was rather theoretical. I found during MBA studies in management that, while I relished my technical background, people interactions were far more interesting.

My first job after graduation took me across the country and exposed me to practical electrical applications. Uncle Sam called me and helped me get more training, but I realized military life was not my preference, especially when so many activities applied negative motivations. Back in civilian life I benefited from a variety of trainings, also discovering that field service work was very reactive and not as appealing to me. I did find diagnosing circuit problems and program managing large installations was a good fit. Shifting into project work and sales development provided the basis for much of my further career, embracing a proactive approach to make things happen.

Working in medical equipment field service and installations for GE gave me appreciation for customer service, and I finally hit my mark when shifting to factory work in computers with HP. Manufacturing, sales development, program management, and finally in a corporate project office, over a 22-year span, was a fabulous training ground for my professional development. During that time, I met up with cultural anthropologist Dr. Robert J. Graham, and we wrote our first book together, *Creating an Environment for Successful Projects*. Becoming a published author was a labor of love, to codify his observations in conjunction with my practitioner experiences. This led to future works such as *Creating the Project Office* and multiple editions, and then to collaborating with fellow practitioner Alfonso Bucero on *The Complete Project Manager* and *Project Sponsorship*.

Having worked in corporate environments for many years, I then made the shift to become an independent consultant and trainer for the Englund Project Management Consultancy. I developed and continue delivering university graduate certificate courses focusing on people skills in project management. I served as a long-term training seminar instructor for the Project Management Institute, first with Dr. Graham and then with Alfonso Bucero. Numerous speaking engagements for PMI global gatherings provided opportunities to share learnings with other searchers. It is very satisfying to help professionals change thinking, explore other mindsets, integrate people skills, and help them create organic environments that bring career advancements and greater fulfillment. I relish every engagement, whether in workshops, consulting engagements, and graduate university courses, to learn from and share insights with others.

WHY TO BELIEVE IN?

People believe in all kinds of things. Some people believe in aliens, while others don't. Some believe in global warming and climate change, while others don't. Some believe in evolution, while others don't. Some believe in God or Gods, while others don't. Some believe in project management and its value and others don't. Why is there this disagreement? Why should we believe in something?

But, before we look at the reasons people do believe things, it might be helpful to consider reasons they might not. It is often said that there are three fundamental reasons people reject or accept a belief. They are: *intellectual, emotional, and moral.* We think this applies in general. See Figure 1.5.

Intellectual: The data, or at least the individual's view or interpretation of the data, is the basis for rejecting a stance. This does not imply that they have a correct understanding of the facts, but it does support their judgment. For instance, if someone is aware of astrophysics and fine-tuning principles, they may reject the notion of aliens.

Emotional: When a decision is made based more on a person's experience and response than on a sound intellectual examination, this is known as an emotional rejection. For instance, someone who doesn't trust environmentalists may oppose global warming.

Moral: When one's will takes precedence over logical analysis, a moral rejection takes place. This does not imply that the choice was incorrect. Without understanding nutrition or scientific research, someone who prefers dark chocolate-covered coffee beans over jellybeans may experience health benefits.

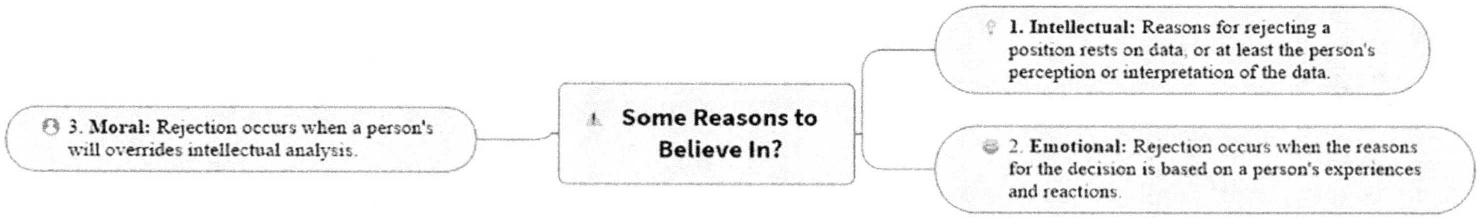

FIGURE 1.5
Some reasons to believe.

We hope it is clear why moral and emotional justifications for beliefs might be problematic. But it's important to remember these. The intellectual will nearly always prevail if those are the causes.

Some Reasons Because of People Believe Things

We found in literature some reasons people believe things, and we summarize them as follows:

First, Societal

What one learns from their culture's common knowledge (conventional wisdom) lends credence to this argument. It is affected by many factors that may or may not be trustworthy. When you go outside in the cold with a wet head, for example, you get sick. Conventional wisdom is not always the best basis for beliefs, even though it is sometimes quite helpful.

Second Reason: Power

What someone in a position of authority, like parents, teachers, coaches, or managers, might tell you serves as the basis for this argument. Your parents may have emphasized that you could drown if you fail in the water and remain there for an extended period of time. Regardless of whether you've ever gone swimming or researched how water affects the lungs, it would probably be wise to trust this. One of the best arguments for believing something is authority. Most of our knowledge is based on this idea.

Third Reason: Psychological

This kind of foundation is predicated on your emotional and psychological state. Examples would be beliefs you hold because they make you feel better or offer you hope. Perhaps since someone is nice to be around, you think they are not lying to you. It is rather simple to understand how such a grounding could be incorrect. However, that does not imply that it is in error.

Fourth Reason: Religious

Authority above has a lot in common with this kind of grounding. But with the addition of otherworldly involvement, it goes a little farther. For instance, a scripture might at least have the potential to be more accurate and knowledgeable if it were revealed by a deity. This grounding may be extremely untrustworthy if the religion is a lie. But this grounding would be superior to anything we can have if the religion is real. Even our understanding of reality is subject to error. The only completely trustworthy source of information would be God, as understood by Christianity or even classical theism, who is all-knowing and incapable of lying.

Fifth Reason: Philosophical
This type of grounding is based on a correspondence of beliefs and facts. It is, in fact, called the Correspondence Theory of Truth. As J. P. Moreland put it:

> *In this simplest form, the correspondence theory of truth says that a proposition is true just in case it corresponds to reality. When what it asserts to be the case is the case ...*
>
> *Grasses being green makes Sally's thought true even if Sally is blind and cannot tell whether or not it is true, and even if Sally does not believe the thought, reality makes thoughts true or false. A thought is not made true by someone believing it or by someone being able to determine whether or not it is true.*

While this Correspondence Theory might seem obvious, it is important to note how this differs from the others. While the others might be grounded in reality, they do not have to be. In other words, they could be correct, but they might not be.

WHY TO BELIEVE IN SOMETHING?

Because it is true! That's the only good reason. The Christian might say, what about the Bible? Yes, the Bible is a special case but not for the reason some might assume. When someone asks you why you believe something, especially Christianity, your first answer should be, "because it is true". Then, back that up by presenting the case from Scripture and evidence.

Since the 17th century, the scientific method has been how scientific knowledge is built and validated. This method is based on an ongoing cycle used to determine and validate conclusions:

1. Ask a question
2. Do some basic research to understand the existing facts
3. Form a hypothesis for the answer to the question
4. Test the hypothesis with an experiment
5. Analyze the results and data from the experiment
6. Report conclusions

This cycle repeats over and over again until final results are achieved. The most effective project managers have learned to apply the scientific method to their projects. The project management scientific method follows the same basic structure but for project plans and work.

1. Set a goal (charter)
2. Conduct initial planning with the project team
3. Publish a baseline project plan
4. Execute the project following the baseline plan
5. Measure progress and evaluate new data, details, information
6. Adjust the project plan and/or charter as needed

A project manager can swiftly produce an initial "good enough" plan that incorporates all the information currently available by using the scientific approach. This first plan is then recorded as a "baseline" or preliminary image. The project manager can then continuously modify the plan to take into account new information that is learned as the project progresses. Does that imply that you will release a plan that isn't accurate? Indeed! Every project plan is created using the most up-to-date information available at the moment. This indicates that most project plans end up being incorrect. The trick is to accept that right away. In actuality, the plan outlines our desired outcome rather than what will actually occur.

A common misconception among project managers is that a project plan cannot be altered. Steps one through four are followed, but they end there. Failure to deliver on a plan after it has been publicized is the project manager's fault. This indicates that the team postpones actually beginning the job and lengthens the project planning cycle. The project team's general stress level decreases when the project manager adopts a scientific method approach to project planning. When planning a project, perfection is neither required nor expected. The goal is simply to get to a plan reflecting the current knowledge to get the work going. If (or when) new information comes up that disproves the hypothesis (initial plan), adjustments can be made, a new baseline set, and the cycle can continue.

Adopt the scientific process and think of your strategy as your best hypothesis right now, whether you're managing a project for your company or for personal objectives. Then, as soon as you can, begin your experiment (the labor for your project) and attempt to validate or refute your hypothesis. After that, modify and repeat.

ARE YOU READY TO BE A BELIEVER?

What can we learn about project management from the scientific method? Your project plan is a hypothesis, to provide you with an idea. Projects can proceed more quickly and with less stress if you adopt a few scientific ideas. In what ways has project management helped you succeed personally? Describe it as an exercise.

Tool
Great. Based on its content, review the questionnaire titled "Why Do You Believe In ...?". It blends the reasons for belief (intellectual, emotional, moral, societal, power, psychological, religious, philosophical) with reflective questions that help participants explore their own stance.

Questionnaire: Why Do You Believe In ...?

Part 1: General Belief Exploration

1. What is something important you believe in strongly (e.g., God, science, project management, global warming, etc.)?
2. Why do you think you believe in it?
    - Intellectual reasons (facts, evidence, logic)
    - Emotional reasons (feelings, experiences, personal impact)
    - Moral reasons (values, preferences, lifestyle choices)

Part 2: Influences on Belief

3. How much do societal influences (culture, traditions, common wisdom) affect your beliefs?
    - Not at all
    - A little
    - Somewhat
    - Very much
4. How much does authority (parents, teachers, experts, leaders) influence what you believe?
    - Not at all
    - A little
    - Somewhat
    - Very much

5. Do your psychological needs (hope, comfort, motivation) shape your beliefs? If yes, how?
6. Do you hold any religious-based beliefs? If yes, what role do they play in your decision-making?
7. Do you agree with the **correspondence theory of truth** ("A belief is true if it matches reality")? Why or why not?

Part 3: Belief and Truth

8. Do you think people should believe in something *because it is true* or because it is *useful/helpful*? Explain your answer.
9. Has there been a moment in your life when you changed a belief after gaining new evidence or perspective? Describe it.
10. In your personal or professional life, how do you test whether a belief is true (e.g., research, experience, authority, faith)?

Part 4: Belief in Project Management (optional adaptation from your document)

11. Do you believe in the value of **project management**? Why or why not?
12. How has project management (or structured planning) helped you succeed personally or professionally?
13. Do you see project management as closer to a science (based on testing and evidence) or an art (based on experience and people)? Explain.

Part 5: Reflection

14. What beliefs help you stay motivated and positive in your daily life?
15. If someone asked do you believe in [X]?, what would your first response be?

2

The Foundation

Believe in yourself! Have faith in your abilities! Without humble but reasonable confidence in your own powers you cannot be successful or happy.

Norman Vincent Peale

BELIEF DEFINITIONS

The term "belief" has multiple definitions depending on the context. Here are definitions:

1. **General Definition**: A belief is acceptance, especially in the absence of evidence that something exists or is true. For instance: A project manager believes that the best policy is to be honest.
2. **Philosophical Definition**: In Epistemology (the study of knowledge) belief is a mental state in which a person holds something to be true, regardless of evidence. For example: Alice believes that the earth orbits the Sun.
3. **Psychological Definition**: A belief is a cognitive depiction of reality that shapes a person's feelings, ideas, and actions. For example: a junior project manager's self-confidence can influence success in the future.
4. **Religious Definition**: A conviction or trust in a higher power or religious doctrine. For example: Christian beliefs emphasize faith, love, and salvation.
5. **Cultural and Social Definition**: Beliefs are shared ideas or assumptions that shape a society's customs, norms, and values. Example: In many indigenous cultures, there is a belief that nature is sacred and living beings are interconnected.

DOI: 10.1201/9781003730736-2

6. **Scientific Definition:** In science, beliefs are considered provisional assumptions or hypotheses that require testing and verification. Scientific beliefs are subject to change based on new evidence. For example: A scientist believes that climate change is caused by human activity based on overwhelming evidence from atmospheric data, climate models, and peer-reviewed research.

"Believe" comes from an old German word meaning "to have faith." Belief does not require proof; acceptance is sufficient. If you are a jury and the prosecution and defense present different versions of what happened, you have to decide which side to believe. It is also possible to think of confidence as belief. Believing in me does not mean that you think I'm real, unless I'm a ghost. It shows you that I can achieve my objectives. Among the synonyms for believe that we might encounter in literature are accept, comprehend, take, trust, buy, assume, credit, swallow, presume, suppose, account, lay store on, conclude, set store by, infer, deduce, and accredit. The words believe and belief are often confused by writers.

Believe means to accept as true, particularly without absolute certainty (i.e., as opposed to knowing) when used as a verb. Belief means mental acceptance of a claim as likely true when used as a noun. A good way to remember the difference is that Believe is a verb. 'Believe' is the most frequently used of the two words. It occurs roughly twice as often as 'belief'. Middle English *bileven*, which comes from Old English, means "to have faith or confidence" (in a person), according to etymonline.com. Midway through the 13th century, the phrase "be persuaded of the truth of" (a doctrine, system, religion, etc.) was coined; early in the 14th century, it was used as "credit upon the grounds of authority or testimony without complete demonstration, accept as true."

GUIDING PRINCIPLES FOR PROJECT MANAGERS

Project managers frequently approach their work and team leadership with a set of values or guiding principles. The following are some typical opinions that many project managers have. See Figure 2.1.

Clear Communication Is Key

Everyone will agree with the project's objectives, schedule, and expectations if there is effective communication. In addition to preventing

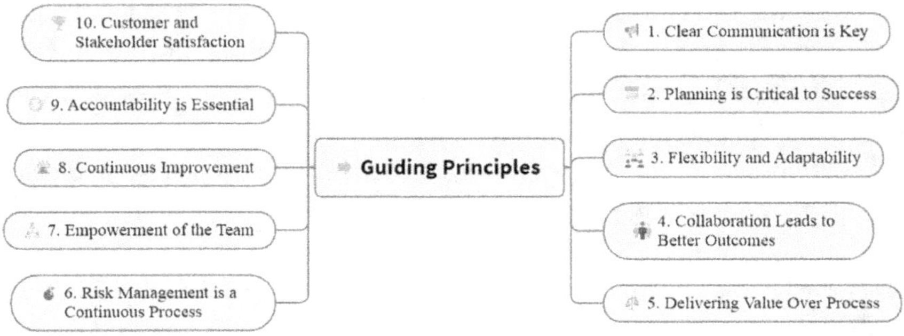

FIGURE 2.1
Guiding principles or set of PM values.

misunderstandings, open communication makes sure that everyone in the team feels informed and supported.

Planning Is Critical to Success

Proper planning—setting clear goals, timelines, and milestones—is crucial. It helps manage risks, anticipate obstacles, and ensures that all stakeholders are on the same page.

Flexibility and Adaptability

Things don't always go according to plan. A good project manager believes in the importance of flexibility and adaptability in responding to changes, whether those changes are related to scope, timelines, or resources.

Collaboration Leads to Better Outcomes

A strong project manager values teamwork and collaboration. They believe that involving the team in decision-making processes leads to more creative solutions, better problem-solving, and a greater sense of ownership.

Delivering Value Over Process

The goal is to deliver value to stakeholders and customers. Project managers often prioritize delivering the right results over rigidly adhering to processes. They understand that flexibility in approach can be necessary to ensure value is achieved.

Risk Management Is a Continuous Process

Project managers believe that identifying, assessing, and managing risks is an ongoing process. They make sure to have strategies in place to mitigate risks, but they also understand that some level of risk is inherent in any project.

Empowerment of the Team

Effective project managers believe in empowering their teams. Giving team members the resources, authority, and responsibility to make decisions fosters engagement, innovation, and accountability.

Continuous Improvement

A good project manager believes in learning from both successes and failures. They encourage the team to constantly reflect on what worked well and what could be improved for future projects.

Accountability Is Essential

Clear expectations and accountability help ensure the team stays focused and meets objectives. A good project manager believes in holding both themselves and their team accountable for meeting deadlines and delivering quality results. Accountability means all persons commit to overall success of the team, project, organization, in addition to their own tasks.

Customer and Stakeholder Satisfaction

The project manager's focus is not only on project success but also on meeting or exceeding stakeholder and customer expectations. They believe in maintaining strong relationships with stakeholders and adjusting the project to meet their needs and requirements.

BELIEVE IN YOUR PROFESSION

Would you say you follow any of these beliefs in your work, or do you have your own guiding principles?

A project manager believes in their profession for several reasons, most of which are tied to the sense of purpose, achieving a vision, responsibility, and impact they can have on both organizations and individuals. Here are some key reasons why project managers are passionate about their work:

Making a Tangible Impact

Project managers often feel a strong sense of accomplishment because they are directly responsible for turning ideas into reality. They oversee the planning, execution, and delivery of projects that can positively impact the organization or society. Whether it's launching a new product, improving a process, or building something new, project managers believe in the tangible outcomes of their work.

Problem-Solving and Overcoming Challenges

Project management is often about finding creative solutions to complex problems. Every project comes with unique challenges—budget constraints, scope changes, resource shortages, team dynamics—and overcoming these challenges is one of the aspects that draws project managers to their profession. They enjoy the satisfaction that comes with solving problems and keeping projects on track.

Continuous Learning and Growth

The project management profession offers continuous opportunities for growth and development. Every project is different, which means project managers are constantly learning, adapting, and honing their skills. This ongoing personal and professional development keeps the role engaging and meaningful.

Leading and Supporting Teams

Project managers have the privilege of leading diverse teams and fostering collaboration. They often believe in the power of teamwork and enjoy seeing team members grow in their roles and achieve success together. The ability to inspire and empower others gives project managers a sense of fulfillment and purpose.

Driving Organizational Success

Project managers believe that they are integral to an organization's success. By managing projects efficiently, they help their companies achieve strategic goals, improve efficiency, and maintain competitiveness. This sense of contributing to the bigger picture and organizational success is a key motivator.

Variety and Dynamic Work

The diversity of projects in a project manager's career means that no two days are ever the same. This variety keeps the profession exciting and prevents it from becoming monotonous. Project managers thrive on the dynamic nature of their role, whether it's tackling a new industry, working with different teams, or managing unique challenges.

Sense of Responsibility

A project manager often has a sense of ownership over the project's outputs and outcome. This responsibility can be both a challenge and a motivator. Being trusted with leading a project, managing resources, and meeting deadlines is something that project managers take pride in. This responsibility also gives them the opportunity to make a real difference in the organization or community.

Achieving Goals and Delivering Results

Project management is all about setting clear goals and achieving them. Project managers believe in their ability to deliver results—whether it's completing a project on time, within budget, or meeting client expectations. The satisfaction of seeing a project through from start to finish is a strong belief that drives many project managers.

Building Relationships and Networking

Successful project managers believe that building relationships with stakeholders, clients, and team members is key to project success. They value the networks they develop and the partnerships that evolve from their work. These relationships often extend beyond individual projects and can be rewarding both professionally and personally. They put fun on the agenda so that people relish the times they spend together.

Making a Difference in People's Lives

Many project managers believe that the projects they manage directly impact people's lives, whether by improving services, creating new products, or enhancing experiences. They take pride in knowing that their work benefits others, whether on a personal or societal level.

Job Security and Demand

The profession of project management is in high demand across a wide range of industries, from construction to technology to healthcare. Many project managers believe in the stability and career growth opportunities that come with the role. Knowing that their skills are highly sought after gives them confidence in the longevity of their career.

Ability to Innovate

Project managers often find themselves at the forefront of innovation, especially in industries like tech or product development. They believe in their ability to drive change, experiment with new ideas, and lead their teams toward innovative solutions.

CONCLUSION

In short, project managers believe in their profession because it allows them to create meaningful change, solve complex problems, and contribute to organizational success. They enjoy the variety and challenge that comes with each project and take pride in leading teams and delivering results. Their role not only provides a sense of personal satisfaction but also allows them to make an impact on the world around them. Believing in your profession is crucial for motivation, fulfillment, and long-term success. Here are some reasons why people might believe in their profession, whether it's project management or any other field. See Figure 2.2.

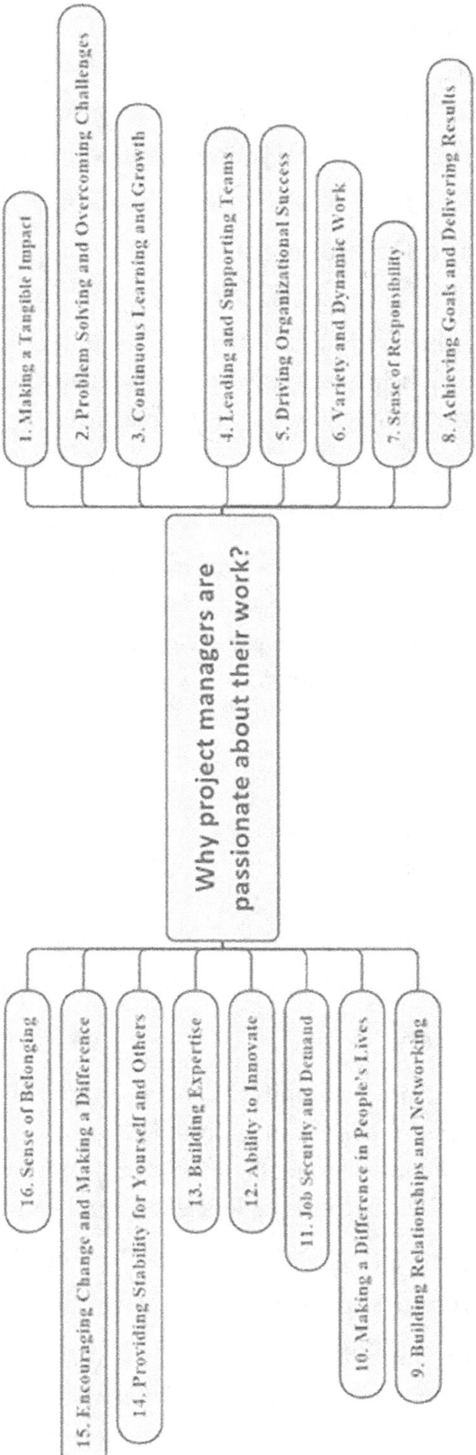

FIGURE 2.2
Why project managers are passionate.

Making a Positive Impact

Every profession offers an opportunity to make a difference. Whether you're solving problems, helping others, or creating something meaningful, believing in the positive outcomes of your work provides purpose and pride.

Personal Growth and Development

A belief in your profession often stems from the opportunities it provides for personal growth. The continuous learning, skill-building, and challenges you face in your career help you grow both professionally and personally. Integrating skills from multiple disciplines is a big plus.

Pride in Your Work

When you believe in your profession, you take pride in what you do. The knowledge that your work is valuable and contributes to a larger goal can be deeply satisfying, whether it's through providing a service, creating something innovative, or supporting a team.

Job Satisfaction

People who believe in their profession are more likely to experience job satisfaction. When you feel connected to what you're doing and see how your role fits into a larger purpose, it boosts motivation and engagement. Believing in your work helps you enjoy the journey, not just the outcome.

Contributing to Society

Many professions have a direct or indirect impact on the well-being of society. Whether you're working in healthcare, education, technology, or social services, knowing that your work benefits others can be a powerful motivator. Create a sense of contributing to the greater good of all concerned.

Career Progression and Stability

A belief in your profession often comes from the opportunities for career advancement. When you trust in your field's potential for growth, whether in terms of skills, responsibilities, or financial rewards, it can provide a sense of security and long-term satisfaction.

Passion and Interest

Believing in your profession often stems from a deep interest or passion for what you do. Whether it's a love for problem-solving, creativity, seeing things getting done, or helping others, a passion for your work fuels perseverance through challenges and enhances your overall job satisfaction.

Building Relationships

Professions often bring people together, whether it's through teamwork, networking, or client interactions. Building lasting, meaningful relationships with colleagues, clients, or mentors can be a key reason to believe in your profession, as these connections can be personally and professionally enriching, often even outlasting the project inself.

A Sense of Purpose

Many people find meaning in their work when they see it as contributing to something greater than themselves. Whether it's helping people, advancing knowledge, or improving systems, having a sense of purpose in your profession can keep you motivated and engaged. Purpose is defined as an enduring reason for a person, team, or group to exist and work together.

Innovation and Creativity

Many careers allow you to innovate, think creatively, and push boundaries. If your profession encourages new ideas, experimentation, or problem-solving, it can be highly fulfilling to know that you are part of something that drives progress and change. Unlock the genius in both yourself and others.

Opportunities to Help Others

In many professions, the ability to directly or indirectly help others—whether it's through providing care, offering advice, or creating solutions—is a key reason to believe in your work. Helping others can bring a sense of joy, purpose, and fulfillment.

A Sense of Achievement

In any profession, achieving milestones or completing important projects brings a sense of accomplishment. Whether it's a successful project, meeting targets, or overcoming challenges, these moments of achievement affirm the value of your profession and the work you do.

Building Expertise

People often believe in their profession because it allows them to become experts in their field. Mastery of skills, knowledge, and processes can create a sense of pride and self-confidence, motivating individuals to continue growing and contributing to their field.

Providing Stability for Yourself and Others

Many people believe in their profession because it allows them to provide for themselves and their families. Financial stability, job security, and career growth are important factors that contribute to believing in the long-term value of your career.

Encouraging Change and Making a Difference

Some people are driven by the desire to enact change. Whether it's changing organizational practices, improving societal systems, or making advancements in a particular field, believing in your profession can mean believing in the potential for positive change and progress. Change thinking, perhaps organically, that leads to improved environments for creating successful working environments.

Sense of Belonging

Being part of a professional community or network can foster a sense of belonging. Sharing values, ideas, and goals with others in your field can help you feel connected to a larger purpose, increasing your belief in your profession. Believing in your profession goes beyond just the tasks and duties you perform—it's about feeling that your work aligns with your values, contributes to your growth, and makes a meaningful impact. This belief fuels your drive, commitment, and passion, helping you navigate challenges and ultimately find satisfaction in what you do.

HOW A PROJECT MANAGER'S BELIEFS CAN IMPACT PROJECT SUCCESS

The belief that a project manager holds about their profession and the approach they take can have a significant impact on the success of a project. A project manager's mindset, values, and leadership style directly influence how they handle challenges, make decisions, and inspire their team. Here are several ways a project manager's beliefs can impact project success:

Influence on Team Morale and Motivation

- **Belief in Team Collaboration:** A project manager who believes in teamwork and collaboration will foster a positive team culture and psychological safety. This, in turn, helps boost morale and motivation, leading to higher productivity and better outcomes.
- **Supportive Leadership:** When a project manager believes in supporting their team and empowering them, it creates an environment where team members feel trusted and valued. This can lead to increased engagement, accountability, and a greater commitment to project goals.
- **Impact on Success:** Teams that feel supported and motivated are more likely to work efficiently, resolve conflicts quickly, have fun, and go above and beyond to ensure project success.

Clear Communication and Alignment

- **Belief in Transparent Communication:** A project manager who values clear, transparent communication ensures that everyone involved in the project understands the goals, expectations, and status of the project at all times.
- **Impact on Success:** This belief helps avoid misunderstandings and misalignment, reducing the likelihood of scope creep or delays. When team

members, stakeholders, and clients are on the same page, projects are more likely to be completed on time and within scope.

Risk Management and Proactive Problem-Solving

- **Belief in Risk Mitigation:** A project manager who believes in the importance of risk management will prioritize identifying potential risks early on, seek to avoid risks when possible, and create mitigation strategies. They will also have contingency plans in place for unforeseen challenges.
- **Impact on Success:** A proactive approach to risk management can prevent small problems from becoming major roadblocks, leading to fewer disruptions and a smoother project execution.

Commitment to Quality

- **Belief in Delivering High-Quality Results:** When a project manager believes that quality is non-negotiable, they ensure that all deliverables meet or exceed all required standards. This may involve thorough testing, frequent quality checks, and constant feedback loops with stakeholders.
- **Impact on Success:** High-quality deliverables lead to greater customer satisfaction and stakeholder approval, which increases the likelihood of long-term project success, repeat business, and positive reputations.

Adaptability and Flexibility

- **Belief in Adaptability:** A project manager who believes in flexibility and adaptability understands that no project ever goes exactly as planned. They will be more open to adjusting project scope, timelines, or methods to accommodate changing circumstances, new information, or unforeseen challenges.
- **Impact on Success:** Projects that encounter unexpected changes are less likely to fall into failure if the project manager is adaptable. They can pivot quickly, reallocate resources, and ensure that the project stays on track despite external pressures or internal challenges.

Focus on Delivering Value

- **Belief in Customer/Stakeholder Value:** A project manager who prioritizes delivering value to customers or stakeholders will focus on ensuring that project outputs and outcomes align with business goals and provide meaningful benefits.
- **Impact on Success:** By keeping the end user or stakeholder needs in focus, the project manager ensures that the project delivers expected results, which leads to increased customer satisfaction, a sense of accomplishment, and a higher probability of project success.

Vision and Strategic Thinking

- **Belief in Alignment with Business Goals:** A project manager who believes in the importance of aligning project goals with the broader organizational strategy, and crafting a vision statement that identifies project success, will ensure that the project supports long-term business objectives, not just immediate deliverables.
- **Impact on Success:** Projects that are aligned with company goals are more likely to be deemed successful, as they contribute directly to the organization's overall growth and success. Strategic thinking, as an individual competency, means seeing the bigger picture and linking all actions, tasks, and projects so they contribute to fulfilling strategic goals.

Managing Time and Resources Effectively

- **Belief in Efficient Resource Allocation:** Project managers who believe in optimizing time and resources ensure that every aspect of the project is planned efficiently, and resources are allocated effectively. They are diligent about managing budgets, schedules, and personnel.
- **Impact on Success:** Efficient resource management leads to reduced costs, minimal delays, and optimal performance, which are key factors in delivering a successful project on time and within budget.

Building Stakeholder Trust

- **Belief in Trust and Transparency with Stakeholders:** A project manager who believes in fostering trust with stakeholders ensures regular

updates, honest assessments of progress, and transparent communication about potential issues or risks.
- **Impact on Success:** When stakeholders trust the project manager, they are more likely to offer support, provide necessary resources, and maintain patience during challenges, which improves the likelihood of project success.

Commitment to Continuous Improvement

- **Belief in Learning and Adaptation:** A project manager who values continuous improvement and actively seeks feedback, evaluates project performance, and implements lessons learned from past experiences ensures that each project becomes an opportunity for growth and refinement.
- **Impact on Success:** By improving processes, methodologies, relationships, and approaches with each project, the project manager increases the efficiency and effectiveness of future projects, contributing to long-term success across the organization.

Decision-Making and Accountability

- **Belief in Accountability:** A project manager who believes in taking ownership of decisions and outcomes will hold themselves and the team accountable for meeting objectives. They address issues head-on and ensure that everyone is clear on their responsibilities and commit to being accountable, not only for their own tasks but for overall success.
- **Impact on Success:** Accountability helps ensure that all team members are focused on their tasks and deadlines, preventing bottlenecks and ensuring that the project progresses smoothly.

CONCLUSION

A project manager's beliefs significantly impact their leadership style, decision-making, and approach to project execution. Positive beliefs—such as valuing teamwork, communication, flexibility, and customer focus—lead to a more effective and successful project. On the other hand, negative or disengaged beliefs can hinder progress, cause delays, and reduce the quality of outcomes. The alignment of a project manager's beliefs with best practices in project management directly influences the likelihood of meeting project objectives and delivering successful results.

BELIEVE COMPONENTS

Believe components refer to the parts of a belief system, such as what a person believes, why they believe it, and how strongly. It is common in psychology or behavior studies. For example, in attitude theory, beliefs are often broken into components like **cognitive** (what you think), **affective** (how you feel), and **behavioral** (how you act). In project management, belief components typically relate to the factors that influence a project team's confidence, motivation, and commitment to success. These components are included in Figure 2.3.

1. **Clarity of Vision**: The team needs to believe in the project's purpose and objectives. A well-defined vision helps align efforts and maintain motivation. Elaborating the project's purpose and objectives takes some time, but it is a worthwhile team effort. We suggest organizing a team discussion meeting to elaborate a project's purpose and vision. If you want to learn more about how to develop your project purpose and vision, read the book *The Complete Project Manager* (Englund & Bucero, 2019).
2. **Leadership Support**: Strong sponsorship and executive backing instill confidence in the team, reinforcing their belief that the project is important and achievable. Executive support is needed for effective project sponsorship and successful projects (Bucero, 2024; Englund & Bucero, 2015). We recommend training executives and project managers together on the sponsorship subject because many executives

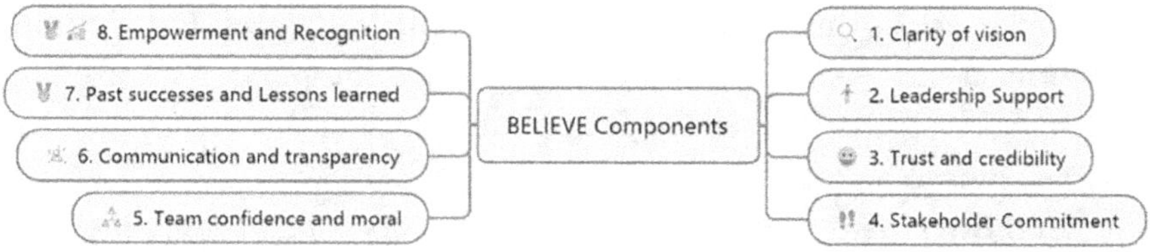

FIGURE 2.3
Believe components.

have not been trained on that matter, and it is essential to work together with executives to achieve successful projects.

3. **Trust and Credibility**: Team members need to believe in their leader's competence and integrity. Trust fosters collaboration and a willingness to overcome challenges. Trust and credibility are key skills to develop by every project manager, and they help to increase their level of influence when managing projects.
4. **Stakeholder Commitment**: If key stakeholders actively support the project, the team is more likely to believe in its success and feel valued. As much as the project manager engages project stakeholders, the project requirements will be more concise and achievable. Achieving stakeholder commitment consists of listening to the music project stakeholders play and empathizing with them.
5. **Team Confidence and Morale**: A positive, engaged team with the right skills and mindset strengthens belief in achieving project goals. Share among your team the sentence "you could be my teacher." Create a positive atmosphere where people are more and more confident with each other. Use positive attitude and reinforce positive behaviors every day.
6. **Communication and Transparency**: Open communication builds trust and ensures that everyone understands the project's progress, challenges, and next steps. The number one step about communication is to listen to your project stakeholders and find out their communication needs and preferred channels for communications. How to do it? Ask them. Be transparent, share current project status, the good and the bad news with your team, and you will create a climate of communication and transparency. Do not kill the messenger.
7. **Past Successes and Lessons Learned**: Belief grows when teams see evidence that similar projects have succeeded, and they can learn from past experiences. It is one of the reasons why running retrospective analysis sessions throughout and at the end of project life cycles are so important. Do not reinvent the wheel and jeopardize project success. The project manager and the project sponsor need to work together to sell the features, benefits, and advantages of those sessions to executives.
8. **Empowerment and Recognition**: When team members feel empowered to contribute and receive recognition for their efforts, their belief in the project and their own abilities increase. We always suggest

giving the benefit of the doubt to your team members and other project stakeholders and you will be surprised by them. In terms of recognition the sentence "I appreciate your contribution, thanks" only takes little saliva from your mouth. It is free of cost.

As a project manager your focus is on inspiring, motivating, and leading people for project success, then reinforcing these belief components could further strengthen your team's commitment and engagement. In project management, belief components typically relate to the factors that influence a project team's confidence, motivation, and commitment to success.

HOW TO ENHANCE ANY OF THESE IN YOUR CURRENT PROJECT

Enhancing your project team's confidence, motivation, and commitment is crucial for achieving project success. Here are several strategies to foster these belief components:

Building Team Confidence

1. **Support Learning and Development**: Offer opportunities for team members to enhance their skills through training, workshops, or conferences. Encouraging continuous learning and sharing experiences boosts their confidence in their roles and contributions.
2. **Celebrate Achievements**: Acknowledge both small and significant accomplishments. Celebrating milestones reinforces the team's belief in their abilities and fosters a positive atmosphere.
3. **Provide Clear Expectations and Necessary Tools**: Ensure that team members have the resources and clarity they need to succeed. Clear instructions and the right tools eliminate ambiguity, allowing them to perform confidently.

Enhancing Motivation

1. **Foster a Collaborative Environment**: Encourage teamwork and open communication. A collaborative atmosphere ensures that everyone feels involved and valued, leading to increased motivation.

2. **Provide Regular Feedback**: Offer constructive feedback regularly. Recognizing achievements and providing guidance for improvement helps team members stay motivated and aligned with project goals.
3. **Recognize and Reward Efforts**: Implement recognition programs to acknowledge individual and team contributions. Public recognition and rewards can significantly boost morale and motivation.

Strengthening Commitment
Align Tasks with Strengths and Interests: Assign tasks that match team members' strengths and interests. This alignment fosters a sense of ownership and commitment to the project's success.

1. **Involve the Team in Decision-Making**: Engage team members in planning and decision-making processes. When they have a say in how work is done, they are more invested and committed to the outcomes.
2. **Create a Positive Work Environment**: Cultivate an atmosphere of trust, respect, and support. A positive environment enhances job satisfaction and strengthens commitment to the team's objectives.

By implementing these strategies, you enhance team confidence, motivation, and commitment, leading to improved performance and successful project outcomes.

Believe Synonyms
Merriam-Webster Dictionary defines believe as: ***to accept*** to regard as right or true: only the most naive car buyer would have *believed* the salesman's claim that the dealership was actually losing money on the deal.
There are some synonyms of the word believe:

- accept
- understand
- take
- trust
- buy
- swallow
- credit
- assume
- set store on

- presume
- suppose
- set store by
- account
- conclude
- infer
- deduce
- accredit

Synonyms and Examples

- **accept**
    - *Most people accept what the newspapers say as being correct.*
- **credit**
    - *It's hard to credit that she's 87.*
- **swallow**
    - informal
    - *I personally find it hard to swallow the official narrative.*
- **buy**
    - informal
    - *When it comes to global warming, he doesn't buy it and is out to discredit the whole theory.*
- **fall for**
    - informal
    - *People are too savvy to fall for that nowadays.*
- **assume**
    - *When she wasn't home, I assumed she was at work.*
- **suppose**
    - *I suppose we have to accept those constraints.*

Tools

Assessment: Strengthening Beliefs for Project Success based on the believe components

This assessment helps measure how well your project environment fosters **confidence, motivation, and commitment** in your team. Respond to each statement on a **1–5 scale**:

(1 = Strongly Disagree, 2 = Disagree, 3 = Neutral, 4 = Agree, 5 = Strongly Agree).

Part 1: Team Confidence

1. I feel I have the necessary skills and opportunities to grow in my role.
2. Our team celebrates both small and major accomplishments.
3. I clearly understand what is expected of me in the project.
4. I have access to the tools and resources needed to perform my tasks well.
5. I believe in our team's ability to achieve project success.

Part 2: Motivation

6. Collaboration and open communication are actively encouraged in our team.
7. I regularly receive constructive feedback that helps me improve.
8. My contributions are recognized and appreciated by the project leader and peers.
9. I feel inspired and motivated to contribute to project goals.
10. Team members support each other, which keeps motivation high.

Part 3: Commitment

11. My tasks are aligned with my strengths and interests.
12. I am involved in decision-making processes that affect my work.
13. I feel ownership of our project's success.
14. The work environment promotes trust, respect, and mutual support.
15. I intend to stay fully engaged until project completion.

Scoring and Interpretation

- Confidence Score (Q1–5): ___ / 25
- Motivation Score (Q6–10): ___ / 25
- Commitment Score (Q11–15): ___ / 25
- Total Score: ___ / 75

Guidelines

- 61–75 → Strong belief environment (team is highly confident, motivated, and committed).
- 46–60 → Moderate belief environment (some areas to improve).
- ≤45 → Weak belief environment (requires immediate attention to strengthen team beliefs).

3

The Value of Belief

Your beliefs become your thoughts, Your thoughts become your words, Your words become your actions, Your actions become your habits, Your habits become your values, Your values become your destiny.

Mahatma Gandhi

THE VALUE OF BELIEF

Each of us is fundamentally different because of our beliefs. They are a significant factor in both peace and conflict; while different views can lead to revolutions, shared beliefs unite people. The power of beliefs to both energize and polarize is more evident than ever in the era of social media, artificial intelligence, and the internet. This brings up an important query: how do people form their opinions?

The premise that people create an internal model of the world to guide their decisions in order to help them achieve external goals, like obtaining rewards and avoiding punishments, is the foundation of a classic method to analyze beliefs (Figure 3.1). Therefore, encourage people to develop accurate views based on all the information they come across that is pertinent to their objectives.

For example, if you read an article that suggests that social distancing is recommended by the Centers for Disease Control (information), you may form a belief that social distancing reduces the spread of disease and keeps your distance from other pedestrians (action selection) to maintain your health (goal).

DOI: 10.1201/9781003730736-3

(A) A traditional approach to studying beliefs is grounded on the idea that the value of a belief is based solely on its ability to guide action selection to optimize external outcomes (e.g., gain food and money, avoid physical harm). Following the delivery of external outcomes, individuals update their beliefs about which rewards/punishments are available, and which actions lead to better outcomes. Individuals may also select actions that are expected to lead to the delivery of sensory information that can help inform future action selection.

(B) According to the revised framework, individuals are also motivated to optimize internal outcomes (e.g., positive emotions, a sense of validation, confidence). Internal outcomes are derived from the beliefs themselves. Thus, individuals are additionally motivated to select actions they expect will lead to the delivery of sensory information that will generate beliefs that lead to desired internal outcomes. Individuals may also update their beliefs in response to external outcomes in ways that make desired beliefs, and thus internal outcomes, more likely.

However, this narrative does not provide the complete tale. There is growing evidence that beliefs are more than merely means to an end. Instead, people are motivated to keep beliefs because they are a source of value in and of themselves. For instance, people typically prefer to think that they are right rather than wrong, that the future is bright rather than gloomy, and that their opinions are certain rather than uncertain. In the most remarkable instances, people deliberately ignore evidence that might contradict a preferred worldview, even when doing so hinders pursuit of external goals, to establish their desired beliefs even at the price of truthfulness.

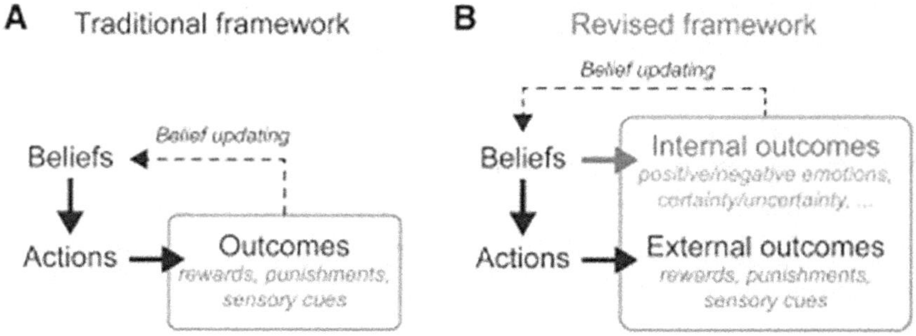

FIGURE 3.1
The value of beliefs.

We recognize that this implies the brain views ideas as a significant source of internal outcomes that are either rewarding or punishing in and of themselves, rather than only as tools to achieve external outcomes (3B). For instance, we may feel validated and experience happy emotions that serve as an internal reward if we think we will definitely receive a grant. However, being unsure of the result might cause anxiety and serve as an internal punishment.

Many disciplines, including psychology, philosophy, and economics, have put forth theories of motivated belief creation (for a summary of this literature, see Sharot and Garrett, 2016). But neuroscientific research has only lately started to reveal how motivation affects the brain's construction of beliefs. With contributions from numerous research groups, this topic has rapidly grown to be extremely busy. Here, we highlight important discoveries from linked human and animal research. We examine the reasons behind the brain's self-motivation to construct beliefs.

Neural Representation of Beliefs

The acknowledgment that a proposition is true is a frequent definition of a "belief." Beliefs can be held with some degree of ambiguity since most beliefs are about hidden states of the world that cannot be immediately observed. This includes, for instance, beliefs about unknowns that are historically distant (like the history of our ancestors), geographically far (like the existence of aliens), or masked by noise (like a fog-covered street sign). Examining how beliefs are formed from noisy sensory experiences is a typical experimental strategy used to research beliefs across species (e.g., Shadlen & Kiani, 2013).

Participants are rewarded for accurately reporting whether a sensory stimulus is in one of two states (e.g., whether a cloud of moving dots has more leftward or rightward motion). Participants therefore learn to convert noisy, ambiguous sensory information into ideas about the hidden state of the world and to base reports on those beliefs in order to maximize reward.

The Value of Belief of a Project Manager

Because belief has a direct impact on stakeholder involvement, team motivation, and overall project success, it is a potent tool for project managers. Here's why belief is so important:

1. Encourages Confidence and Motivation: A project manager who genuinely believes in the project's success encourages their team to remain

dedicated and resilient. A project manager with a pleasant attitude, for instance, inspires positive behavior in their team and all other project stakeholders.

- When challenges arise, belief keeps the team focused on solutions rather than setbacks. And it is the project manger's obligation to work on that.

2. Strengthens Leadership Influence:
When a project manager exhibits confidence and conviction, stakeholders and executives are more inclined to support the initiative. In order to favorably impact all project stakeholders, project managers need self-confidence. Belief enables the project manager to convince leadership of the project's worth.
3. Builds a High-Trust Team Culture
    - Teams take cues from their leader—if a project manager believes in the team's capabilities, members feel more empowered and motivated. An example is giving team members the benefit of the doubt.
    - It fosters a growth mindset where challenges are seen as opportunities for improvement rather than failures.
    - Projects often face unexpected risks and obstacles, but belief helps a project manager stay proactive and optimistic. This mindset prevents panic, enabling clearer decision-making under pressure.
5. Enhances Stakeholder Engagement
    - When a project manager is passionate about the project, it is simpler to win over sponsors and clients. This idea lowers resistance and fosters more cooperation by bringing all parties together around a shared objective.
6. Encourages Innovation and Problem-Solving
    - A project manager who believes in the potential for success encourages creativity and risk-taking, leading to innovative solutions.
    - This belief creates a safe environment where team members feel confident to propose new ideas.

CASE STUDY

Several years ago, I (Bucero) had the initiative to create a local PMI association (PMI Barcelona Chapter) in the city of Barcelona (Spain). As a PMI member since 1992, I was a PMI Madrid Chapter founder, but I could not be part of the PMI Madrid Board because of different issues. Then I decided to create a PMI Chapter in Barcelona to better serve the needs of local PMI members in the Catalunya Community.

I was born in Madrid and historically Madrid and Barcelona had some type of rivalry, but I defended that the project management profession has no borders. So, I proceeded as follows:

1. I formed a team composed of two professionals from Madrid and the rest of the team from Barcelona inspired confidence among them about that project was feasible and possible if we acted as a team. All team members felt motivated about that challenging project. I traveled to Barcelona once per month with my colleague Javier Sanz (Chapter Secretary) to have our monthly meeting, and we did it for almost two years.
2. I was confident in myself because I had been in the profession for many years and I showed my passion for that project. That behavior achieved 25 professional followers and with the PMI Global support we chartered that PMI Chapter in September 2004. Two years later I gave up my President position and asked a Barcelona colleague to apply for that role. I was aware that I was not a Catalunya guy, and they had the right to be managed by one of them.
3. One of our successes was that the Board of Directors of that Chapter built up a lot of confidence and support among the team, and it has been one of the main heritages from its initiation up to date. As our secretary and I lived in Madrid, we delegated many tasks to Barcelona board members, and we worked very well as a team.
4. We started up a web page and ran monthly meetings to serve better our members as a method to enhance stakeholder engagement. Inviting professionals to share their project war stories in front of the rest of the team members periodically.
5. That PMI Barcelona Chapter creation was not a path of roses. I found a lot of obstacles because initially PMI Madrid Chapter saw us as their competitors in Spain, when our objective was to spread out the profession in Spain and share best practices with them.

Another case study comes from noted educator and risk specialist Carl Pritchard:

> Our organization was pioneering in the e-learning space, working to serve a major university and a telecommunications client in the mid-1990's. I had laid out a vision as to where we could go and how we could address client needs over a 56K modem. I believed we could meet the breadth of what these clients wanted and needed, despite the (by today's standards) primitive technology at our disposal. A vendor came in to pitch a different approach to my boss, and I stood firm that their approach was not client-centric and was going to create more headaches than it would solve. Over my objections, my employer opted for the external vendor, rather than doing the work in-house. Weeks went by with little progress, and our clients became increasingly frustrated with the relationship. Over time, the vendor proved to be unable to deliver, and the client deadlines loomed large. My employer asked if I still believed we could succeed with the approach I had initially laid out, and I affirmed we could. The vendor was dropped. With a small team of skilled professionals and the most rudimentary tools at our disposal, we generated our company's first successful forays into distance learning and retained both of our critical clients. The vendor ultimately went out of business. As a 30-something, I had been tempted to give up on my vision. Instead, I believed in what we originally laid out, and built infrastructure that lasted for years (while the technology caught up).

Michael O'Brochta, a good friend and colleague provides his example:

> I had a "light bulb" moment many years ago early in my career as an electrical engineering project manager at the CIA. After a couple of early career successes, I was selected to attend an intensive two-week training session that included several small group exercises. Each exercise was facilitated by a physiologist.
>
> Unbeknownst to me at the time, the small group I was assigned to consisted of people who thought/believed/acted much differently than I did. In hindsight, I realized that my assignment was based on the extensive pre-training assessment I had completed. I had been intentionally assigned to work with people different from me. Our small group was given hypothetical problems to solve that centered on mountain climbing. Since I had recently climbed one of the world's tallest mountains, I immediately identified with the problems. None of the other members of the small group had any knowledge whatsoever about mountain climbing. I was the expert. I actually scoffed at the idea that any of the others could contribute to the problem solutions.

The small group toiled away at the problems and repeatedly arrived at solutions that did not fully rely on my expertise. As the expert, I considered those solutions to be substandard. However, I needed to be open minded. In the end, I had to admit that the small group solutions were better than mine, even though I was the expert. Really. Ever since, I cherish opportunities to be around others who are different from me. I am open minded. My go-to question when working with others is, "What do you think?"

ADDING VALUE

We are sure you heard sentences like: "We want our project managers to add value." "How are you adding value to the organization?" "I want to spend more time on valued-added activities." But what does adding value mean?

We found several things we think you can do to add value (in a meaningful way) as a project manager:

Team Building

Projects are done by people. People make up teams. Groups of disparate people don't have the same impact as a well-functioning team. Therefore, spending time on team building is worthwhile and will create value for the organization because you'll be better at delivering whatever it is you are delivering. Focus on creating a positive work environment. Think about what people need to get their tasks done. Look for roadblocks you can remove, processes you can streamline. Talk about the blockers and why they are a problem. And get some fun in there, too. Examples of project team building activities tailored to enhance collaboration, communication, and problem-solving within your project team are:

Escape Room Challenge

- **Objective:** Enhance problem-solving and teamwork under pressure.
- **How It Works:** Teams are "locked" in a themed room and must solve puzzles to escape within a set time.
- **Benefits:** Fosters collaboration and quick thinking (teambuildingworld.com).

SWOT Analysis Workshop

- **Objective:** Develop strategic thinking and decision-making skills.
- **How It Works:** Teams analyze a project or challenge by identifying its Strengths, Weaknesses, Opportunities, and Threats.
- **Benefits:** Encourages critical thinking and strategic planning.

Scavenger Hunt

- **Objective:** Promote teamwork and problem-solving in a fun setting.
- **How It Works**: Teams follow clues to find items or complete tasks within a time limit.
- **Benefits**: Enhances communication and collaboration (Vantage Circle).

Egg Drop Challenge

- **Objective:** Encourage creativity and engineering skills.
- **How It Works:** Teams design a contraption to protect an egg from a high drop using limited materials.
- **Benefits:** Stimulates innovation and teamwork.

Charades with a Twist

- **Objective**: Improve non-verbal communication and team bonding.
- **How It Works**: Team members act out words or phrases related to the project or industry without speaking.
- **Benefits:** Enhances understanding and rapport among team members.

Engineering Challenge

- **Objective:** Develop problem-solving and collaboration skills.
- **How It Works:** Teams use provided materials to build a structure or solve an engineering problem.
- **Benefits:** Encourages practical application of skills and teamwork.

Blind Square

- **Objective:** Enhance communication and trust.

- **How It Works:** Team members, blindfolded, must work together to form a perfect square using a rope.
- **Benefits:** Builds trust and improves verbal communication skills.

Hackathon

- **Objective**: Foster innovation and cross-functional collaboration.
- **How It Works**: Teams work intensively on a project or problem over a short period, often outside their usual tasks.
- **Benefits:** Encourages creativity and team cohesion.

Office Olympics

- **Objective**: Promote friendly competition and team spirit.
- **How It Works**: Teams compete in a series of fun, office-related challenges.
- **Benefits**: Boosts morale and encourages teamwork.

Cooking Class

- **Objective:** Enhance collaboration and creativity.
- **How It Works:** Teams work together to prepare a meal under guidance.
- **Benefits:** Fosters communication and teamwork in a relaxed environment.

Tenacity

Being dedicated to the team, the task, and the project is a surefire method to provide value since it raises the likelihood that the project will be completed. How many initiatives do you know of that got off the ground but lacked the necessary momentum to succeed? Tenacity will help you prevent that. Ensuring your time pays off for the business depends on your capacity to follow through and complete the work, assuming you are working on the proper projects. In a project team, exhibiting tenacity entails persistent resolve, resiliency, and a shared dedication to conquering obstacles. The following striking instances demonstrate team perseverance in a variety of settings:

Healthcare Transformation in Nairobi
At Aga Khan University Hospital in Nairobi, Kenya, the implementation of point-of-care (POC) testing faced significant challenges due to inconsistent processes and equipment. A cross-functional team, led by the laboratory quality coordinator, embarked on a comprehensive overhaul. Their persistent efforts resulted in a fivefold reduction in medical errors and a 70% increase in timely responses to critical alerts, demonstrating remarkable tenacity in improving patient care.

Process Improvement in a Stagnant Organization
In a project where initial resistance from senior leadership threatened progress, a determined team formed under a development manager's guidance. Despite challenges, they collaborated across departments to enhance project management, development, and QA processes. Their relentless pursuit of improvement led to significant advancements, showcasing the power of tenacity in transforming organizational practices (projecttimes.com).

Apollo 13 Mission Crisis Management
During the Apollo 13 mission, a critical failure in the spacecraft's service module jeopardized the lives of astronauts. A specialized "Tiger Team" of engineers and scientists, led by NASA's Gene Kranz, worked tirelessly under immense pressure to devise a safe return plan. Their unwavering determination and collaborative problem-solving were instrumental in bringing the crew back safely, exemplifying exceptional team tenacity (Wikipedia).

Innovative Product Development Under Adversity
A project team faced multiple setbacks, including design failures, manufacturing challenges, and funding delays. Rather than succumbing to adversity, the team exhibited tenacity by reworking designs, securing alternative manufacturers, and presenting compelling cases to obtain necessary funding. Their perseverance culminated in the successful launch of a new product, albeit slightly behind schedule, highlighting the importance of resilience in achieving project goals (career support).

Triumph Over Adversity in Sports
Chris Nikic, the first person with Down syndrome to complete an Ironman triathlon, embodies personal tenacity. His journey began with modest goals, gradually building up to the rigorous demands of an Ironman. His mantra of "1% better" and consistent effort, supported by his father and coach,

underscores the power of incremental progress and determination in overcoming challenges.

These examples illustrate that tenacity within project teams is characterized by persistence, adaptability, and a collective commitment to overcoming obstacles. Whether in healthcare, organizational change, space exploration, product development, or personal achievement, tenacious teams demonstrate that challenges can be transformed into opportunities for growth and success.

Relationship-Building

This is such a broad topic, encompassing dispute resolution, overcoming awkwardness, being diplomatic while speaking truth to power, respectful challenge, and understanding who to connect with and when. This category includes a wide range of soft talents, sometimes known as power skills these days. They are crucial because they enable you to complete tasks even in challenging environments.

You will comprehend more and find it easier to complete your tasks if you listen more. You'll have a better understanding of the business environment, which will enable you to make the best choices that, you guessed it, result in a higher-value outcome. To promote trust, cooperation, and a healthy work environment, a project team needs to establish strong relationships.

In the political arena, when facing tough economic headwinds, leaders should be telling the truth—pointing out signs of progress but acknowledging areas where that progress is slower. While stakeholders don't expect things to get better overnight, messages that do not ring true for many who are unable to reconcile that messaging with the realities in their own lives may increase worries that there is no plan to address their concerns.

The following are some useful exercises and tactics to improve team member relationships:

Coffee Chats

- **Objective**: Facilitate informal conversations to strengthen interpersonal connections.
- **How It Works**: Pair team members randomly for a casual coffee break, either in person or virtually. This allows individuals to share personal interests and experiences, promoting mutual understanding and camaraderie.

- **Benefits**: Encourages open communication and helps break down hierarchical barriers.

Scavenger Hunt

- **Objective:** Promote teamwork and problem-solving in a fun setting.
- **How It Works:** Organize a scavenger hunt with clues that require collaboration to solve. Teams work together to find items or complete tasks within a set time frame.
- **Benefits:** Enhances communication and fosters a sense of unity among team members.

Role Reversal Scenarios

- **Objective:** Develop empathy and understanding of colleagues' challenges.
- **How It Works:** Assign team members to take on each other's roles for a day or specific tasks. This experience provides insight into the responsibilities and pressures faced by others.
- **Benefits:** Builds appreciation and respect for diverse roles within the team (weberprojectmanagement.com).

Improv Workshop

- **Objective**: Enhance creativity and adaptability in communication.
- **How It Works**: Conduct an improv session where team members participate in spontaneous acting exercises. This encourages quick thinking and supports a non-judgmental environment.
- **Benefits:** Improves communication skills and fosters a collaborative spirit.

Outdoor Yoga Session

- **Objective**: Promote relaxation and mindfulness among team members.
- **How It Works:** Arrange a group yoga session in a tranquil outdoor setting, guided by a certified instructor.
- **Benefits:** Reduces stress and strengthens team bonds through shared wellness activities.

💬 Story Swap

- **Objective:** Enhance active listening and empathy within the team.
- **How It Works**: Pair team members to share personal stories or experiences, with one person speaking while the other listens attentively. Afterward, discuss insights gained from each other's narratives.
- **Benefits:** Deepens understanding and connection between team members.

Goal Setting Workshops

- **Objective**: Align individual aspirations with team objectives.
- **How It Works:** Facilitate a workshop where team members collaboratively set personal and professional goals, discussing how these align with the project's success.
- **Benefits:** Encourages accountability and mutual support within the team.

Control the Process

Although it may not seem like a very worthwhile endeavor, understanding and utilizing project management procedures makes it much easier to organize, standardize, save time, automate, compare, and improve. Everyone will know what to do and what to anticipate if there is a standard approach, even if it is informal. This helps to reduce uncertainty during what is typically an uncertain time for people—projects to deliver change come with an overhead of having to live with not knowing exactly what the future will look like. The team and the larger stakeholder group may experience additional worry and stress as a result. Ensure processes are not overwhelming, are streamlined and understandable.

Change Management

Projects start to feel out of control when change is not managed appropriately, and that's when stakeholders start to get nervous. Create and get agreement on a change management process early in the project life cycle. You can help your projects be more successful and 'land' better with the receiving organization, if you manage change properly. That goes for both the process-led effort of receiving and handling change requests as part of project management work and also integrating what you are delivering into the business in a way that makes it possible for the benefits to be received as soon as possible, with the least disruption. Benefits = value. How do you interpret 'adding value' as a project manager? It could go much further than what is written here. There are many other ways of looking at roles and how to serve stakeholder communities in the most value-adding way.

HOW CAN A PROJECT MANAGER ADD VALUE TO YOUR PEERS?

Your responsibilities as a project manager go beyond managing assignments and deadlines; by encouraging teamwork, increasing productivity, and promoting project success, you play a crucial role in bringing value to your colleagues. Here's how you can have a big influence:

Facilitate Effective Communication

It is essential to have open and transparent channels of communication. All team members are kept informed and in sync by regular updates, attentive listening, and open information exchange. Real-time collaboration can be streamlined by using programs like Microsoft Teams or Slack. Eliminate the benefits of poor communications to individuals and promote good communications that keep projects running smoothly.

Empower Your Team

Acknowledge and capitalize on each team member's strengths. Encourage motivation and a sense of ownership by giving activities that complement each person's skills and offering chances for professional growth. Increased productivity and job satisfaction result from this empowerment.

Promote a Collaborative Culture

Encourage teamwork through joint decision-making and shared problem-solving. Involving peers in decisions not only enhances engagement but also brings diverse perspectives that can lead to innovative solutions. Organizing team-building activities can further strengthen relationships and trust within the team.

Manage Stakeholder Relationships

Understanding stakeholder expectations, keeping lines of communication open, and quickly resolving issues are all necessary for effective stakeholder management. Guarantee alignment and support, which are essential for project success, by serving as a liaison between a team and stakeholders.

Implement Risk Management Strategies

Proactively identifying potential risks and developing mitigation plans can prevent issues from escalating. By addressing risks early, you protect the project's scope, budget, and timeline, ensuring smoother execution.

Foster a Positive Work Environment

Effective stakeholder management requires understanding stakeholder expectations, maintaining open channels of communication, and promptly resolving difficulties. By acting as a liaison between a team and stakeholders ensures alignment and support, both of which are critical for project success.

Lead by Example

Setting an example for peers involves being dedicated, flexible, and solution-focused.

Approachability and readiness help others boost self-esteem and motivate the team to take initiative. Concentrating on these areas not only improves the effectiveness of projects but also supports peers' professional development and contentment. Teamwork and leadership foster a productive and happy workplace.

MEASURING THE VALUE

To show how project management affects organizational success, its value needs to be measured. To evaluate this value, the Project Management

Institute (PMI) provides a number of metrics and methods that include both quantitative and qualitative elements.

Performance Metrics

These metrics evaluate how effectively a project is meeting its objectives. Key indicators include: (HogoNext)

- Schedule Performance Index (SPI): Measures the efficiency of time utilization in a project. An SPI greater than 1 indicates the project is ahead of schedule.
- Cost Performance Index (CPI): Assesses cost efficiency. A CPI greater than 1 suggests the project is under budget (HogoNext).
- Earned Value (EV): Represents the value of work actually performed, aiding in assessing project performance against the plan.

These metrics help in identifying trends and making informed decisions to steer projects towards success (www.pmi.org).

Benefit-Cost Ratio (B/C Ratio)

This financial metric compares the benefits derived from project management to its associated costs. A higher ratio indicates greater value. However, it may overlook intangible benefits, necessitating a balanced approach in evaluation.

Return on Investment (ROI)

ROI calculates the profitability of a project by comparing net benefits to total costs. A positive ROI signifies that the project delivers more value than its cost, highlighting the effectiveness of project management practices (pmiuk.co.uk). Note that project activity is negative value compared to positive outcomes.

Intangible Benefits

Not all value can be quantified financially. Intangible benefits such as improved customer satisfaction, enhanced team morale, and better stakeholder relationships play a crucial role in overall value assessment (pmiuk.co.uk). Intangible benefits of a project manager, while not always directly measurable in numbers, significantly impact project success and organizational health. Here are some key intangible benefits along with suggestions on how they *can* be indirectly measured or observed to assess the project manager's value:

- *Team Morale and Engagement*
    - **Benefit**: A good project manager boosts team morale and fosters engagement.
    - **How to Measure**:
        - Employee satisfaction surveys
        - Team turnover rate
        - Feedback from 1:1 meetings or retrospectives
- *Stakeholder Trust and Confidence*
    - **Benefit:** Builds trust with executives, sponsors, and clients.
    - **How to Measure:**
        - Stakeholder feedback or Net Promoter Score (NPS)
        - Frequency and quality of stakeholder involvement
        - Willingness to assign critical projects
        - Repeat business
- *Conflict Resolution and Relationship Building*
    - **Benefit:** Helps manage and resolve conflicts, maintaining a healthy team dynamic.
    - **How to Measure:**
        - Number of unresolved conflicts escalated
        - Peer feedback
        - Observations during collaboration or crisis moments
- *Cultural Alignment and Positive Attitude*
    - **Benefit:** Reinforces company values and cultivates a positive culture.
    - **How to Measure:**
        - Cultural audit results
        - Observation of team behavior (e.g., collaboration, resilience)
        - Pulse surveys

Adaptability and Problem-Solving

- **Benefit:** Anticipates and adjusts to change quickly, keeping the project on track.
- **How to Measure:**
    - Response time and effectiveness during change requests or risks
    - Project recovery success rate
    - Innovation or process improvements suggested
    - Defects or issues resolved rates

Mentorship and Team Growth

- **Benefit:** Develops others, leading to skill growth and internal promotions.
- **How to Measure:**
  - Career progression of team members
  - Skills development metrics
  - Internal feedback loops on mentoring
  - Willingness to work together again

Communication and Influence

- **Benefit:** Translates complexity and influences decisions.
- **How to Measure:**
  - Feedback from executives and team members
  - Effectiveness of decision-making meetings
  - Clarity and completeness of project communications

Reputation and Leadership Brand

- **Benefit:** Enhances the organizations and team's credibility.
- **How to Measure:**
  - Repeat project assignments
  - Recognition or awards
  - Invitations to participate in strategic planning

Project Management Maturity
Assessing an organization's project management maturity level provides insights into its capability to manage projects effectively. Higher maturity levels often correlate with better project outcomes and more consistent value delivery (pmi.org).

Customer Satisfaction
Feedback from clients and stakeholders is a direct indicator of the perceived value of project management efforts. High levels of satisfaction often reflect successful project delivery and effective management practices.

Conclusion

A thorough assessment of the value of project management combines financial analysis, qualitative evaluations, and performance measures. Organizations can obtain a comprehensive picture of how project management affects their success by combining these methods.

Tool
Here's a structured **assessment on the value of belief** that you can use for reflection, workshops, or team evaluations:

Assessment: The Value of Belief

Purpose

This assessment helps individuals and project managers reflect on how belief influences motivation, leadership, collaboration, and project success.

Section 1 – Personal Belief and Motivation

1. I believe in my own capacity to achieve success, even when facing obstacles.
   - Strongly Disagree | Disagree | Neutral | Agree | Strongly Agree

2. My beliefs help me stay focused on solutions rather than setbacks.
   - Strongly Disagree | Disagree | Neutral | Agree | Strongly Agree

3. When I truly believe in a goal, I feel more motivated to inspire others.
   - Strongly Disagree | Disagree | Neutral | Agree | Strongly Agree

Section 2 – Leadership and Influence

4. My belief in the project's success positively influences executives and stakeholders.
   - Strongly Disagree | Disagree | Neutral | Agree | Strongly Agree

5. I communicate with conviction and passion when I believe in the project's value.
   - Strongly Disagree | Disagree | Neutral | Agree | Strongly Agree

6. My belief in the team's capabilities fosters empowerment and trust.
   - Strongly Disagree | Disagree | Neutral | Agree | Strongly Agree

Section 3 – Team and Stakeholder Engagement

7. I show my belief in team members by giving them the benefit of the doubt.
   - Strongly Disagree | Disagree | Neutral | Agree | Strongly Agree

8. My belief helps me maintain optimism and clear decision-making under pressure.
   - Strongly Disagree | Disagree | Neutral | Agree | Strongly Agree

9. When I believe in the project's purpose, I can align stakeholders toward a common goal.
   - Strongly Disagree | Disagree | Neutral | Agree | Strongly Agree

Section 4 – Belief and Innovation

10. I encourage creativity and innovation because I believe in the team's potential for success.
    - Strongly Disagree | Disagree | Neutral | Agree | Strongly Agree

11. My belief helps me create a safe environment for team members to share new ideas.
    - Strongly Disagree | Disagree | Neutral | Agree | Strongly Agree

Section 5 – Reflection on Value

12. My beliefs bring me positive emotions (validation, confidence, hope).
    - Strongly Disagree | Disagree | Neutral | Agree | Strongly Agree

13. I recognize that belief is not only a tool for external goals but also an internal source of motivation.
    - Strongly Disagree | Disagree | Neutral | Agree | Strongly Agree

14. I can identify moments in my career where belief directly added value to project outcomes.
    - Yes | No | Not Sure

Scoring and Reflection

- **Mostly Strongly Agree/Agree:** You view belief as a powerful driver of motivation, leadership, and value creation.
- **Mostly Neutral:** You may not yet fully recognize the role of belief in your leadership approach. Reflect on situations where stronger belief could have influenced outcomes.
- **Mostly Disagree/Strongly Disagree:** You may rely more on external tools and processes. Consider how cultivating belief could strengthen team trust and stakeholder engagement.

4

Do You Believe?

Choose a job you love, and you will never have to work a day in your life.
Confucius

HOW MUCH DO YOU BELIEVE?

We believe in the project management profession, and we have demonstrated that throughout our professional careers by contributing as practitioners, authors, volunteers in Congresses, and instructors. We may not be sure how much we believe. The problem is how to assess it? In this chapter we try to answer some questions about believing in yourself, in your organization.

Connie Plowman, PMP, is an experienced practitioner, executive, consultant, Gallup-Certified Strengths Coach and Certified in teaching Emotional Intelligence, who reinforces these ideas with her testimonial:

Do you believe? If yes, how do you know? As an education leader, corporate leader, volunteer leader, and project leader, I have found the answer – over the span of my career. As project managers, we have many tools in our PM Tool Kit. To help find the answer, I have tapped into my two favorite tools: knowing what I naturally do best (strengths) and how I show up (emotional intelligence). Believing in myself means understanding what strengths I bring to the team and how I can best contribute as an integral member of the team. It means that I strive to intentionally apply my strengths – every day – which builds confidence. I show up using my self-management skills by managing my emotions and behaviors toward a positive result for myself.

Believing in my team means I leverage the collective strengths of each team member. Maximizing the team's collective talents and strengths leads to a

 DOI: 10.1201/9781003730736-4

high-performance team, which builds trust. I show up using my social awareness skills to recognize the emotions and behaviors of others by observing, listening, and asking questions. Believing in my organization means that my core values are aligned with the organization's core values. Seeing the organization's values in action can build strong relationships and ongoing support. I show up using my relationship management skills to better understand needs – my needs and the organization's needs – and to manage and leverage interactions across the organization.

Believing in my project means that I am passionate about what I do. To echo the authors' statement in the Preface, we do our best work when we love what we do. Being passionate about the project can lead to higher-quality deliverables that results in project success. I show up using my self-awareness skills by controlling my reactions and being comfortable with discomfort, when challenges occur.

I believe in myself … to build confidence. I believe in my team … to build trust. I believe in my organization … to build relationships and support. I believe in my project … to build success. Then ask yourself: What are your strengths? How do you show up? What tools are in your PM Tool Kit? These questions may help you find the answer to, "Do I believe?" They have for me."

BELIEVING IN YOURSELF

The level of success you see in your life is a direct result of your belief system. We're not talking about religion or philosophy here. What we mean is your belief in your ability to succeed. Believing in yourself means having faith in your own capabilities. It means believing that you CAN do something, you can manage your project successfully—that it is within your ability.

Marge Combe, PMP, experienced professional and PMI Fellow, said:

I had an experience in my career that really formed my way to believe in myself. I had been tapped by an executive at my company to take on a new role that had never been done before. After many weeks in the job, I was completely unable to come up with a plan to fulfill the mission of the role. I did not feel I had the skills needed to do it. A colleague asked me a perfect question: what is it that you do especially well, for which you are appreciated? When I answered that question, he said, "Start there, even if you don't yet know where it will take you. It will give you a boost to realize that you can make progress just by doing what you do best." I will never forget that. When I am stuck or feel under-confident, I try to figure out how to adapt what I do best to the needs of the situation. It doesn't have to be the perfect answer. It just has to feel like movement. And because it's something people appreciate about me, it builds my confidence as well as teamwork with others.

When you believe in yourself, you can overcome self-doubt and have the confidence to take action and get things done. When you're drowning in fears, doubts, and self-sabotaging behaviors, success feels out of your grasp. All the skills, training, and tools in the world won't change your life.

The Steps to Believe in Yourself

Here are some steps to develop your belief in yourself:

Work on Your Strengths (Not Weaknesses)

When you fail over and over at something that seems easy to others, it's nearly impossible to believe in yourself. Struggling with confidence, you tend to focus on things you can't do. That's because you feel weaknesses more keenly. They are painfully highlighted in your mind, symbols of shame, weakness, and failure.

"I'm bad at this," quickly escalates to, "I'll never be good at anything!" The good news is that everyone has weaknesses—and strengths. You need to determine how to identify your strengths, so that you can get the most mileage out of them. The solution to start building confidence right away: Stop spinning your wheels at things you aren't wired for. Find what you are good at already, and work at becoming great. A common habit of successful people is to focus on the positive—what they excel at—and delegate weaknesses to others instead of worrying about not measuring up. When you shift your focus to developing your strengths, you will effortlessly feel more competent and confident. And when you put effort into developing your natural abilities, you can become a "badass." Maybe you already know some of your strengths.

One example is the improvement I (Bucero) felt in dealing with executives on projects over the years. As I was educated in an autocratic culture, strict respect for elderly people was key in my education. I had a lot of respect for my father, and for many years I believed I should deal with executives as though they were my father. Working closer with them helped me to learn that executives are people like me, but they are measured by different measures than I am. It is not unrespectful to discuss logical arguments to be supported during a project. To get some ideas, think about how people might describe you:

- Always has a grand vision, big-picture thinking
- Notices little things, detail-oriented

- Loves learning and research
- Good communicator
- Never gives up
- Hands-on and action-oriented

Talk to your family, close friends, and project management colleagues about what your strengths are. What are some things they come to you for help with? You're not fishing for compliments, though sometimes this exercise can come with a confidence boost. If you don't know what your strengths are, use an assessment tool. For example, you may use "StrengthsFinder" by Tom Rath to pinpoint their natural ways of thinking and behaving. The exercise takes about an hour, walking through a series of questions. It's designed to uncover your hidden sources of power—your strongest qualities. The assessment also reveals some of the best ways you can start using your inborn talents.

Knowing your strengths will allow you to put more of your effort into areas you're already talented in. You can grow in the ways you're naturally inclined. Once you are working from a base of strength and belief in your abilities, you can find resources to tackle anything. Recognizing your natural abilities (and inabilities) gives you permission to stop banging your head against a wall. You can do what you do best and find workarounds for the rest. For example, starting a business requires many skills, but you don't have to own them all:

- You're great at marketing but terrible with numbers. Outsource your bookkeeping and hire a professional to manage the budget.
- You shine at creating content but dread interacting on social media. Use automation tools to share your work or hire a strategist.
- You love big picture planning but get stuck when it's time to implement. Take on a partner who can get your idea off the ground.
- You may discover strengths you have but did not think about, just because they come easy and take less energy.

Use your strengths to your advantage. This approach will bring out your confidence as you focus on using your best skills. Shifting everything else off your plate gives you the power to be awesome at what you do best. Your strengths are what make you unique. Develop these qualities and depend on them.

Be Your Own Coach

If you don't believe in yourself, you won't believe someone who is cheering you on from the sidelines. That's why a great small business coach leads their clients to success not by cheerleading, but by developing talent. According to Gallup's research, great performance is a result of managers who empower employees. Managers who focus on developing talent get outstanding results. Great coaches and managers lead by empowering people to succeed with the right tools, education, and resources to become excellent. They interact with people using approaches like Conversational Intelligence to put people at ease and bring out the best in everyone.

The research shows that positive thinking, goal setting, and performance reviews don't create results on their own. When you don't believe you have what it takes, none of those things are a magic bullet. So how can you take advantage of this knowledge, and coach yourself?

Seek out the tools and education to develop your talent. Take action. Each step you take, moving closer to your goal, is evidence of your ability to succeed. Believing in yourself will become easier and easier the more you see your success reflected back to you. The secret is, you don't need another person to work through this process. You don't need to possess qualities of leadership to coach yourself through it. The trick is to learn more about yourself without being judgmental. Start by journaling on where you want to be in life. Write out all the things you want to have, and the kind of person you want to be. From this, you can come up with strategies to get to that point.

What does success look like for you?

Choosing ideas that suit you—and working with your strengths—can propel you towards your own vision of success. You can also write about the parts of your life that you're unhappy with. Create action steps, and find the resources to improve any of these areas:

Feeling overwhelmed with responsibilities
Wanting more purpose in life
Craving more free time for things you love (reading, gardening, hiking)
Stress about finances

By identifying the parts of your life that you're least satisfied with and keeping an eye on where you ultimately want to be, create long-term goals that are manageable for you. Set one big goal at a time and break it down into smaller parts. Succeeding in taking each small action step, no matter how tiny, leads to visible progress. Ultimately, these consistent small successes add up to self-belief and massive confidence. Over time, you start to see that you can achieve anything you put your mind to—because you are already doing it.

Embrace Who You Are

How can you have faith in yourself when you don't know who you really are? Or worse, you are trying hard to be someone you're not. Self-confidence comes from embracing who you are and what's important to you. It does not come from being inauthentic or trying to impress people.

It is easier said than done. The pressure to fit in, to "be normal," is strong, and it starts early. That's fine if you want to live an unremarkable life. But you're here, reading this book, because that's not life for you. You want something more. To start believing you can have the life you want, dig deep to discover what that will look like for you. Understand what makes you unique and celebrate those things. When you begin living true to your identity and core values, you will start believing in your worth, your abilities, and your human potential. Core values are the foundation for forming your personal philosophy to guide your actions going forward. Embracing yourself is best done in baby steps:

Start by Writing Down What's Important to You

You'll want to do this several times over weeks, or even months. Each time you'll get closer to uncovering your core values. This will help you to see through the things you've been conditioned to believe in. When you're finished, you'll have a list of traits that represent your truest self.

Observe Your Patterns of Thought and Behavior

Do you often cater to others at the expense of your own values? When you think you're being nice or keeping peace, ask yourself why. The end result is that you dull your own shine and live for other people's approval. When you feel this internal conflict happening, work on addressing it. You can kindly say no, assert a different opinion, or let someone know that what they've done has hurt you. Some people might not like the "new you," and that's okay. They can play a more limited role in your life. Fortunately, when you're authentically you, the right people will be drawn to you, and real friendships will grow even stronger.

Step Out from the Expectations of Others

Being authentic, we mean "saying what you believe" (Englund & Graham). Don't be afraid of criticism. Remember that your experience of life is about you, and their experience of life is about them. Doing things that scare you will solidify the confidence you need to be your truest self. It sounds counter-intuitive, but being vulnerable—accepting your fear and not hiding it—is the surest path to building strength. Make time for adventures that are completely out of your normal routine. It could be jumping out of a plane, taking a fitness class, or picking up a paintbrush. Pick something that seems fun but scares you at least a little bit and jump in. You can learn so much about yourself from these activities that you should engage in them regularly. Add variety and do new things often.

Believe You Can and You Will

Your belief is the single most powerful tool you have. You can completely change your life just by changing your beliefs. It's not wishful thinking—believing something changes the way you see the world and yourself. You either see possibilities, or insurmountable obstacles. Which will you choose?

To start believing in yourself, stop believing you are stuck with the skills and abilities you have right now. This is called a fixed mindset, which is a limited thinking pattern that's fatal to your success. Instead, start believing you can change. This way of thinking is called a growth mindset. It means that you believe change is possible. It may be true that you don't have what it takes to accomplish your goals—right now. But if you believe you can improve, develop, and learn, then you will get there—even when nothing goes right. You will grow.

Deeply believing that you and your life can change for the better is the driving force behind all your efforts. You will then be more willing, and even excited, to put in the work required to make those changes. There is no greater confidence booster than seeing the results of your work. But you must believe it's worthwhile or you will never put in the effort. In her TED Talk, Meg Jay tells young people to forget about their identity crisis and start building identity capital. Don't waste another second thinking:

- I'm not qualified for anything.
- I don't know what direction to take.
- I'm not getting anywhere in my life.
- I thought I'd be someone by now.

Building identity capital means doing things that add value to who you are. Start investing time and effort in becoming that "someone" you want to be. I think this advice applies to everyone, even if your 20's are long gone. In fact, it's entirely possible to start a new career at 40, and it's never too late to refine your identity. If you don't have confidence in your skills or abilities, even your appearance, get to work improving:

- Explore new work or educational opportunities.
- Learn new skills or improve the ones you already have.
- Make new connections to uncover new opportunities.

By believing that change is possible, you'll be consciously choosing ways to grow. There is nothing stopping you. Whatever your dream is, once you wholeheartedly believe that it's possible for you, that's when the adventure truly begins.

Be Uncomfortable

If what you did today looks a lot like what you did yesterday, and last week, and three years ago—you're too comfortable. You're in a pattern and it's getting you exactly nowhere. To make real change in your life, get uncomfortable. Step out of your cozy box and do things differently. This means making a greater effort and even feeling a bit weird doing things that are out of the ordinary.

Do Weird Things Anyway

You have this whole life to experiment with who you are. It's your life and you don't need to live it to anyone's expectations but your own. If you want a different life, make it yourself. Through experimenting and acting on your ideas, your capabilities and potential are reflected back at you. Experimenting with different skills, approaches, and tactics in your life is one of the best ways to build confidence and foster mental toughness. Sure, it's uncomfortable! New things always are. Fortunately, when you start to do more new things, the fear becomes fun. You get to learn so much from trying new things, whether you succeed in the attempt or not.

There are challenges. Part of the process means getting over negative self-talk, fear of failure, and imposter syndrome. You may have to face genuine anxiety and work on ways to manage it. Every single person who has made any kind of self-improvement will tell you that these feelings are very real. They'll tell you it's scary, but the results are worth it. Each time you step into something new there will be fear. Your power lies in walking through that fear and doing the thing anyway.

The Power of Believing You can Improve

The magic happens by simply believing it's possible. Your belief in possibility is necessary in work, experimentation, and the consistency needed to change your life. This belief in possibility is what provided rock stars with the single-minded intensity to get on stage every night, in the face of ridiculous odds. It's what gave famous authors the tenacity to keep writing and editing after countless rejected manuscripts. Jack London is a great example of this changed approach. I (Bucero), when writing a new book, not only am looking for knowledge sources, but am testing with other colleagues if my writing may add value to readers, and I reread it several times before sending it to the editor. It's what kept star athletes training, through pain, injury, and loss, until they made it to the top. Now you have some tools to start believing in yourself. Work on these steps consistently, and you will begin to see amazing things happening in your life.

BELIEVING IN YOUR TEAM

The first step is to believe in the person. The second is to not believe their story. What it means to believe in the person is that in your mind, in your gut, you believe that this person can grow, that they can change that, and that they can be better. If that belief is not truly in you, your heart won't be in the coaching and development efforts that you make. And how often do we make the mistake of judging someone too quickly—of evaluating their capacity based on what they can do today instead of seeing the potential in them of what they can become?

When you believe in someone, when you believe in what they can become, it changes how you see them. It changes how you talk to them. And that belief can begin to infect them and generate confidence that they need to take some risky steps to help them grow.

But as you believe in them, be very careful that you don't believe their story. All of us go through tough times, or we have a stuck story. It's a place that we can't get past because we've accepted limits or we've made excuses—and those things are very real to us. We believe them. We think they're true and they become part of our story of why we're not able to go to the next level. For instance, one example is delegation. Delegation begins by determining all the tasks that must be performed to reach project goals. Then select the individual or individuals best qualified to handle each duty and empower them to do it. Finally, check results regularly to make sure the productivity goals you have set are being reached or surpassed.

Why people don't delegate—these are the excuses some European project managers give for not delegating: My team members lack the experience, says J. M. D. (project leader at Spain); it takes more time to explain than to do the job myself; a mistake by a team member could be costly for my project; my position enables me to get quicker action; there are some things that I shouldn't delegate to anyone; my team members are specialists, and they lack the overall knowledge that many decisions require. Other comments gathered after talking to project professionals from the south of Europe were my people are already too busy; my team members just aren't ready to accept more responsibility; I am concerned about lack of control over people's performance when I delegate; I like keeping busy and making my own decisions; delegating is terrifying to me. Is it a problem of insecurity? It seems that the most common reason for failure to delegate is deep insecurity. This self-defeating attitude influences how you accept and recognize performance of those who work under you.

If this is your problem, ask yourself these questions: Is my team member really after my job? Am I afraid he/she will do the work better than I can? Am I afraid to give him/her either overpraise or more authority because I fear that person will replace me? If this is your attitude, it's holding you back. Failing to share authority and responsibility may protect your immediate status at the expense of an opportunity to move up. Delegating some of your authority only makes your work easier.

I (Englund) was planning an internal conference and asked my coordinator if she would negotiate an important contract. She said she'd never done that before but was willing to try. I could coach her along the way if necessary. Little did I know what a giant was unleased through this assignment. I could do the job as I'd done it before but preferred to focus on other aspects of the project. So, I could do other tasks suited to my responsibilities, and she performed marvelously, developing a negotiating skill set that was undoubtedly dormant. It actually opened up a new career path for her in event management.

When you delegate, don't think of it as doing the other person a favor. The use of effective delegation will pay off for you, as project manager and your organization in many ways. You will mobilize resources to achieve more results than you ever thought possible. You will have more time to manage your project, like planning, monitoring team members, and handling conflicts and personnel problems, that no one else can do. You will focus on doing a few tasks very well, rather than doing a lot rather poorly.

You will increase your managerial potential. You will have someone trained to succeed you, so you will not be shackled to one particular area. The organization benefits in many ways. Output goes up, project work may be completed more efficiently, and team members feel free to offer new ideas and suggestions to improve the operation of the company. Decision-making is improved, so the organization becomes more responsive and thus more competitive in the marketplace.

There is more to delegating than saying, "Do what I tell you to do." It is entrusting the matter to the other person. It involves mutual commitment. The person to whom you are delegating makes a commitment to meet your expectations. For you, it means a commitment to give staff full cooperation, backing, and recognition. We suggest following four steps for effective delegation:

1. Define the purpose and importance of the project, together with its deadline and the scope of the delegate's responsibility. The responsibility for clarification lies with you. Don't expect your team members to ask enough questions to clarify this. If possible, get your team members to set their own deadlines. It is usually more effective to delegate a whole job. People feel a greater sense of responsibility when handling an entire job.
2. Provide the necessary authority, resources, and support. Make sure your team member has the authority needed to complete the task. Otherwise, the staffer's requests to others for help and information may be ignored because they don't come from you.
3. Review progress and follow up. Setting deadlines and enforcing them establish a commitment, ensuring their decisions are made promptly and tasks are handled with dispatch.
4. Slowly reduce supervision. On a first assignment, give the person as much supervision as you feel is required. Reduce it on the second assignment if the person was quick and efficient in completing the first task. This will save you time and increase the abilities of the person

handling the task. Encourage all who work for you to increase their duties and responsibilities.

Watch out! Once you have assigned a task, don't be manipulated into relieving the subordinate of the responsibility for taking the next step. Here is how that can work: Barry meets you in the hallway and says, "Alfonso, we have got a problem." Saying the wrong thing, such as "Let me get back to you in an hour," Barry has put the weight back on your shoulders. Instead, see to it that the next move is his. "You are right. There is a problem. Give me a call tomorrow and tell me how it should be handled."

Selecting the right people: one of the most crucial steps is selecting the right person for jobs. Consider factors such as: Friction (disagreement between you and the person taking the assignment is healthy while the assignment is being done. It is only a problem if it extends into the execution stage); Track record (match the job to the person. Past performance is significant only as it relates to the job you are delegating); Location (don't delegate just because someone is close by and convenient to use, or may just not be busy); Organization level (if you want to delegate a job to someone several levels down in the organization, first confer with his or her supervisor and explain the situation. It will make the delegation smoother); Compatibility (ideally, the styles of both persons involved are complementary). Availability is not a skill set.

To get delegated to work, the persons to whom you delegate must have trust in you. Talk is not enough to create this. Be open with your people. Don't try to hide your errors. Provide any information about the organization that would be helpful. As you are working with your team members, clarify their expectations so they know in advance what is expected of them. Treat them with courtesy and respect.

As a project manager, one of your most important jobs is communicating while delegating. Effective delegation demands effective communication. Without it, assignments are blurred, deadlines are vague, and results are predictably poor. Communications calls for mutual consultation and agreement between the manager and team members. It calls, too, for you to solicit team members' reactions and ideas throughout the process.

Start by telling team members very clearly what they are being asked to do. Evoke a feeling of obligation and commitment within team members and then establish a system to reward outstanding performances. During the communicating process, focus on results rather than ways to accomplish the task. That way, team members will learn more, take more initiative, and have greater enthusiasm for the task.

As a follow-up to any assignment, appraise the completed task and discuss your evaluation with team members. This provides insight for both of you, showing how each performed, as well as suggesting how you worked together as a team. A suggestion at this time is, ask yourself two questions: was the task completed as intended in a timely manner? What could be done to delegate a task better in the future?

Discuss your evaluation of their work with team members. Make this a two-way exchange of perceptions and feedback on each other's work style. Give team members positive and negative comments so they understand what they are doing well and where they need to improve. If you want a team's exceptional performance to continue, recognize and reward it. This is a simple, fundamental rule of management. The word "thank you" has a cost of "zero dollars."

If you believe and accept someone else's stuck story, you can't help them. You can't develop or grow that person any further because you've accepted the same limitation. The same excuse that they have made themselves. So, you can't believe their story no matter how true it sounds, and no matter how factual. "Well, there's this obstacle … there's that … or this is my situation, and so that's why it can't happen."

Those are all part of their story, and you absolutely cannot believe their story. The moment you believe the story, you stop being able to coach or develop them any further. Walking that path of believing in them, but not believing them, is the mental key. It's the mental discipline that enables you to be a good coach and develop someone else's ability.

The final mental discipline required for coaching others is staying curious. Curiosity is not just a nice-to-have trait for coaching. It is essential. Let us explain why. When you lose your curiosity, you begin to make assumptions, and assumptions can utterly kill the coaching process. When you coach others, it's probably because you have more age, experience, skill, or wisdom in that area. And that very advantage can become your biggest blind spot. You can jump ahead, finish the person's sentences, and read into the situation before you really have enough information. Then your coaching could push them in the wrong direction.

Even if you are right, the absence of curiosity will be communicated to your coachee as an arrogance or disinterest that will reduce your influence with them. On the other hand, if you cultivate curiosity in yourself, you can model that same approach to your coachee, which is a huge win. If they stop making assumptions themselves and become curious, they open the door to whole new levels of growth and development that were formerly closed to them. Just remember, you are not a coach because you have all the answers. You are a coach because you know the right questions to ask. Your job is not to give great advice and tell them what to do, but rather to help them see the challenge correctly and clearly recognize their own thinking and behavior so that their new choices will be authentic and integrated. It will make sense to them.

Note that It's almost impossible to change other people's beliefs. It all has to start with you. You can affect other's behaviors, however, through your personal examples and coaching.

Here is further advice from a trusted colleague:

> Adopting the best project management mindset, in true Peter Taylor '*Lazy Project Manager*' style, means focusing your energy where it actually counts and resisting the urge to manage everything, everywhere, all at once. It's about being 'productively lazy', understanding the 20% of activities that deliver 80% of the value, and putting your attention there, where it truly counts.
>
> A great project manager doesn't micromanage, instead they orchestrate, they build smart structures, trust their team, and intervene only when their involvement genuinely moves the project forward. This mindset champions preparation over perspiration, clarity over chaos, and calm, well-timed action over constant noise and interference.
>
> And as Peter Taylor often reminds us, the secret isn't doing more, oh no, it is doing the right things but doing them brilliantly and giving yourself (and your team) the space to work smarter, not harder.

BELIEVING IN YOUR ORGANIZATION

In the project management context, believing in your organization means having confidence, alignment, and commitment to the organization's:

- Vision and mission
- Values and culture

- Strategic goals
- Leadership and direction

Here's what it looks like in practice:

Alignment with Organizational Strategy

You understand how your project contributes to the organization's bigger picture, and you communicate that to your team.

→ *We're not just delivering features; we're helping our company become a market leader.*

Trust in Leadership

You trust that decisions made by executives or sponsors are in the best interest of the organization—and your project—even when times are tough.

→ *This doesn't mean blind obedience, but constructive loyalty.*

Advocacy and Representation

You act as an ambassador of the organization when working with external stakeholders, showing pride in the company you represent.

→ *Your credibility as a project manager strengthens when you represent the organization with conviction.*

Commitment and Resilience

When challenges arise, your belief in the organization gives you motivation to persist, adapt, and lead with a positive mindset.

→ *You're more likely to inspire your team if you genuinely believe in where the organization is headed.*

Modeling and Influence

You set the tone for your team: if you believe in the organization, your team is more likely to feel purpose and pride in their work.

→ *Belief is contagious—it boosts morale and trust during uncertainty.*

WHY IT MATTERS

- Projects often fail due to lack of executive support or unclear vision.
- When a PM believes in the organization, they naturally advocate for alignment, persistence, and engagement—all which raise the odds of success.

Would you like a checklist to assess and strengthen this belief within your own project team?

BELIEVING IN YOUR PROJECTS

Project managers have different beliefs based on their experiences, industries, and leadership styles. Here are some common ones:

1. People-Centered Beliefs
   - Success depends on teamwork: The project team is the key driver of success, and fostering collaboration is essential.
   - Stakeholder engagement is crucial: Managing stakeholders effectively determines project success.
   - Positive leadership matters: A motivated and inspired team will perform better.
2. Process-Oriented Beliefs
   - Strong planning prevents failure: A well-structured plan is the foundation of a successful project.
   - Risk management is key: Identifying and mitigating risks early saves time and resources.
   - Following methodology leads to success: Using frameworks like agile, waterfall, or hybrid approaches ensures efficiency.
3. Results-Driven Beliefs
   - Meeting objectives define success: The project is successful if it meets scope, time, cost, and quality requirements.
   - Deliverables matter more than process: The end product is what counts, even if some steps need to be adjusted.
   - Speed is essential: Completing projects quickly ensures competitive advantage.

4. Flexible and Adaptive Beliefs
    - Change is inevitable: Projects need to be flexible, so they adapt to shifting priorities and business needs.
    - Learning is continuous: Every project provides lessons to improve future projects.
    - Failure is part of innovation: Taking risks and learning from failures drives long-term success.
5. Control-Focused Beliefs
    - Strict governance ensures success: Strong oversight, documentation, and compliance prevent issues.
    - Data-driven decision-making is best: Metrics and analytics provide better insights than intuition.
    - Budgets need to be tightly controlled: Financial discipline is essential for sustainability.

Here is a testimonial shared by an experienced practitioner, Becky Winston:

Believing in myself has allowed me to conquer many obstacles. These obstacles ranged from believing I could be a project manager to believing I could be a consultant in project management and be a contributor to other project manager career goals.

I can point to two elements that assisted me to achieve. One was having several moments of self-evaluation that led to self-awareness. This process sounds easy but the dedication to self is not. I have to acknowledge my weaknesses, strengths, opportunities for growth, and the barriers I am facing—both environmental and self-imposed. I try to do this self-evaluation when I have experienced great success and when I have failed.

This evaluation is not done in isolation. I need to take critiques. None of the negative critiques are easy to hear, but if I believe in myself and my ability to adapt and change, I can achieve. To help in determining what is constructive positive and negative criticism, I ask myself if it provides me with a path for self-improvement. If it does, suddenly even negative criticism becomes positive and reaffirms elements of my self-belief.

One caveat is that I do not have to stay in situations that negatively impact my belief in myself. A negative environment is disruptive and feeds aggression and self-doubt over a sustained period of time without providing a path forward for self-improvement or improvement to one's environment. Negative environments breed negative thoughts, and those thoughts will reflect on the people I interact with and in my ability to succeed. These negative environments will also hold me back from seeing what else is available to me to grow and continue to succeed.

Having left environments that are continually shutting doors for advancement have informed my career path. I observed negative treatment of others. I find disapproval if I have engaged in the same behaviors that I view as unethical or unprofessional or are negatively competitive. If I find self-focus, -growth, -reward, and -healing in the decision to extricate myself from a situation, the step is worth the loss of security both professional and financial. These actions have provided me with more self-awareness and the ability to confirm my dedication to the profession.

These decisions are made after self-reflection and meditation, whether several sessions of mindfulness or deep meditation. These somatic quieting tools allow me to dig deeper into analyzing the criticism, do self-evaluation, and develop a professional plan of action. The confidence gained from developing a plan of action and implementing it for professional advancement is truly motivating and rewarding. I remind myself often that self-growth is the value gained. The various benefits of these actions, I continue to share with those project and program managers I mentor. When I see them begin or continue to believe in themselves while continuing to learn, share, and enable others, I am achieving my goals.

Rebecca A. Winston (Becky), JD, President, CPM, Fellow, PMI, rebecca.winston@mycpm.org

Another expert project manager, instructor, and senior consultant, Neil Whitten, PMP, shared with us his experiences about "believing in yourself":

Believing in Yourself: A Cornerstone of Leadership: *I have long been an advocate of believing in yourself—a principle that has shaped my career and the careers of countless project managers I've mentored. This mindset has been my anchor in high-pressure situations, especially when conducting reviews of troubled projects.*

When I first began reviewing troubled projects, the challenges were intimidating. Imagine walking into a room of 10-20 people—project team members, managers, and other leaders—knowing less about the project than anyone there. My credibility was on the line. Every question I asked and every observation I made would either build trust or erode it.

The pressure was intense. Some participants feared embarrassment and became defensive. I faced a wide range of personalities, all under tight constraints. And yet, I had to lead with confidence.

How did I manage? By believing in myself and preparing thoroughly. I set clear expectations, gave participants prep homework, and asked probing questions during the review. Over time, I developed guiding principles that became my foundation:

- *Think for yourself. Rely on analysis, not opinions.*
- *Never avoid necessary confrontation. Address issues despite resistance.*
- *Don't make it personal. Stay objective and professional.*
- *Practice boldness and courage. Pursue concerns with confidence.*
- *Value your own opinion above others'. Stay focused on your mission.*
- *Embrace integrity. Always take the high road.*
- *Trust but verify. Validate every claim with evidence.*
- *Treat others with respect. Lead with empathy.*
- *Think like a leader. Your leadership sets the tone.*
- *Seek mentors when needed. Guidance accelerates growth.*
- *Evaluate yourself daily. Continuous improvement is key.*

The more I practiced these principles, the more natural they became. I wasn't the smartest person in the room—but I was disciplined, prepared, and committed to doing the right thing. Believing in myself opened doors and created opportunities. It can do the same for you.

(Note: Portions extracted from *Power Skills that Lead to Exceptional Performance* by Neal Whitten, published by Business Expert Press)

We provide a tool that you may use to assess your beliefs.

Tool

Based on the contents of this chapter "Do You Believe?," here is a comprehensive survey This survey is designed to be reflective and can be used for personal development, coaching, or within organizations to assess mindset and cultural alignment.

SURVEY: DO YOU BELIEVE?

Instructions

Please answer the following questions honestly. Use the scale provided for each item or answer open-ended where indicated. Your answers will help assess belief levels and identify growth opportunities.

Part 1: Believing in Yourself

On a scale from 1 (Strongly Disagree) to 5 (Strongly Agree), rate the following statements:

1. I believe in my ability to succeed, even when challenges arise.
2. I focus more on developing my strengths than fixing my weaknesses.
3. I actively seek tools, education, or resources to develop my talents.
4. I feel confident in who I am and stay true to my core values.
5. I frequently step out of my comfort zone to try new things.
6. I believe that I can improve any area of my life through effort and learning.
7. I coach myself by reflecting, journaling, and setting personal goals.

Open-ended:

8. What is one thing you currently struggle to believe about yourself?
9. Describe a recent moment when you felt truly confident in your abilities.

Part 2: Believing in Others

On a scale from 1 (Strongly Disagree) to 5 (Strongly Agree):

10. I believe that people can grow and change with the right support.
11. I am careful not to accept others' limiting stories as truth.
12. I maintain curiosity when coaching or helping others.
13. I ask questions rather than assume I know what's best for someone.
14. I see the potential in others, even if they don't see it yet.

Open-ended:

15. How do you help others believe in themselves?
16. What is a limiting belief you've heard someone express that you didn't accept?

Part 3: Believing in Your Organization

On a scale from 1 (Strongly Disagree) to 5 (Strongly Agree):

17. I understand and align with my organization's vision and mission.
18. I trust the leadership team to make decisions that serve the organization's best interests.
19. I represent my organization with pride and professionalism.

20. I remain resilient and motivated even when my organization faces difficulties.
21. I believe my organization fosters a positive, purpose-driven culture.

Open-ended:

22. What does your organization do well that builds your belief in it?
23. What could your organization improve to strengthen trust and alignment?

Part 4: Believing in Your Projects

On a scale from 1 (Strongly Disagree) to 5 (Strongly Agree):

24. I believe that project success depends primarily on the people involved.
25. I trust that thorough planning and risk management lead to better outcomes.
26. I value learning from every project, regardless of its outcome.
27. I believe change and adaptation are essential parts of any successful project.
28. I rely on data and governance to maintain control and clarity.
29. I see failure as a necessary part of growth and innovation.

Open-ended:

30. What core belief do you hold about managing successful projects?
31. What project experience most challenged your belief system—and how did you respond?

FINAL REFLECTION

32. What does "believing" mean to you in the context of your personal and professional life?
33. If you could change one belief today to positively impact your life, what would it be?

5

Developing Your Belief

You only truly believe that which moves you to action.

Douglas Cheney

DEVELOP YOUR BELIEFS

We, as project practitioners, have developed our beliefs based on our professional lives and how project management added value to ourselves, our projects, and our organizations. We have some suggestions to make to develop and strengthen it by:

• KNOWING YOUR "WHY"

Ask yourself why you are a project manager. Are you able to find any argument that can sustain your facts? In some cases, we observed organizations where the project manager position was an individual reward for efficiency, rather than a career aspiration for the professional. Based on our practical experience, we suggest taking some actions that worked well for us:

- Reflect on *why* you chose to become a project manager (or any other profession).
- Connect with the purpose behind your work—whether it's delivering value, enabling teams to succeed, or making a difference in people's lives. Identify your professional purpose.

DOI: 10.1201/9781003730736-5

- A strong why fuels belief. I (Bucero) was assigned as a project manager by accident but after my first project I needed to ask myself why I wanted to continue being a project manager. My main reason was because I like dealing with people and creating a positive environment to succeed with them as a project team. Serving as a leader encourages me to help people to be successful as a team.

  Later on, I identified my purpose that is "helping people and organizations to improving their attitude to achieve project success." My "why" is what inspires me every day to start my day telling my team "Today is a great day and tomorrow will be better." I (Englund) found I perform better being proactive rather than reactive.

• CELEBRATING YOUR WINS

Every professional needs to get recognition for the efforts they make, and it is crucial to celebrate their achievements. We recommend these actions to take based on our practitioner experiences:

- Track successes, lessons learned, and progress over time. For example, we suggest running retrospective analysis sessions at every project milestone. Those sessions basically identify what worked well in the project, what didn't, and what can be learned from that.
- Keep a "success journal" or a "project win wall" to remind yourself and your team of what you've accomplished.
- Reflecting on past achievements reinforces confidence. Sometimes managers focus more on failures than successes, so we suggest imagining yourself wearing a crown on your head and viewing successful projects.

• INVESTING IN CONTINUOUS LEARNING

We, as project management believers, defend the sentence "you could be my teacher." We mean that every project manager can learn from every project stakeholder, focusing on the daily details during project life cycles. I (Bucero) consider myself a continuous learner, and some people criticized me when I introduced myself by saying, "I am Spanish, nobody is perfect." I spend some time every week reflecting upon the skills I need to improve. I (Englund) believe every day is a good day when I learn something new and have fun.

We have these suggestions for you to invest in continuous learning:

- Read books, attend workshops, and learn from other professionals (mentors, peers, or coaches). I (Bucero) usually dedicate at least one hour weekly to read articles or book chapters. I (Englund) spend the first part of each morning reading and reflecting.
- Learn from mistakes, not fear them. Each challenge overcome adds to your belief. Every time you complete a project milestone, do a retrospective analysis and identify lessons learned. This is a wonderful practice to continuously apply.
- Become an integrated thinker, drawing upon multiple disciplines.

• DEVELOPING EMOTIONAL INTELLIGENCE (EQ)

In today's rapidly changing project environments, technical skills alone are not enough to drive success. Project managers also need to possess strong emotional intelligence (EQ) to navigate the complexities of stakeholder relationships, team dynamics, and personal stressors. In our book, *The Complete Project Manager, 2nd Ed* (Englund & Bucero, 2019) and *Toolkit*, we include a survey to assess your emotional intelligence. Our suggestions about how to develop your emotional intelligence are as follows:

- Understand your emotions and how they affect others:

Emotions are powerful forces that guide our thoughts, actions, and relationships. Yet many people move through life without fully understanding what they feel or how those feelings influence others. Learning to recognize and manage your emotions is a key part of emotional intelligence, and it can transform how you connect with the people around you. First, it's important to identify what you're feeling. Emotions like anger, fear, or sadness often hide behind behaviors such as irritability or withdrawal. By pausing to ask yourself, "What's really going on inside me?" you become more aware of the root cause of your reactions. This self-awareness gives you control—and helps prevent emotional responses from taking over. Second, remember that your emotions ripple outward. When you're upset, others may absorb that tension; when you're calm and optimistic, you can lift the mood of an entire group. Being mindful of your emotional energy allows you to communicate more effectively and build trust with others. Finally, empathy ties it all together. Understanding your own emotions helps you recognize similar feelings in others, creating deeper, more compassionate interactions. In short, understanding your emotions isn't just self-care—it's people care. The more you regulate your inner world, the more positive impact you'll have on the world around you.

- Stay calm under pressure and inspire calmness in your team:

In project management, pressure is not an occasional visitor; it's a permanent resident. Deadlines tighten, priorities shift, and unexpected problems surface at the worst possible moments. *The Complete Project Manager* (Englund & Bucero, 2019) and *Today Is a Great Day* books point to a powerful truth: calm is not a personality trait—it's a leadership skill. And when a leader masters it, the entire team benefits. Staying calm under pressure starts with perspective. Effective project managers distinguish between urgency and importance. Not every issue is a crisis, and treating everything like one drains energy and clouds judgment. By pausing, assessing facts, and choosing responses deliberately, leaders prevent emotional reactions from turning manageable problems into major disruptions. Calm thinking creates better decisions—and better outcomes.

Equally important is self-management. Stress is contagious, but so is composure. When leaders regulate their emotions, they send a silent but strong message: we can handle this. Simple habits—slowing your breathing, clarifying next steps, focusing on what can be controlled—help maintain steadiness even when uncertainty is high. As *Today Is a Great Day* (Bucero, 2024) emphasizes, mindset shapes the day. Choosing calm is choosing effectiveness. Inspiring calmness in a team requires visibility and consistency. Teams watch their leaders closely, especially during difficult moments. Clear communication, steady tone, and realistic optimism help people feel safe and focused. Instead of amplifying fear, strong leaders name challenges honestly while reinforcing confidence in the team's ability to respond. Calm leadership doesn't deny problems—it frames them as solvable.

Trust also plays a central role. When expectations are clear and people feel supported, pressure feels lighter. Empowering team members, avoiding blame, and encouraging collaboration reduce anxiety and increase ownership. A calm environment allows creativity and problem-solving to thrive, even under tight constraints. Calm leadership is a strategic advantage. Projects move faster when energy isn't wasted on panic. Teams perform better when they feel grounded. By staying calm under pressure and modeling that calm for others, leaders don't just manage projects, they build resilient, confident teams ready to handle whatever comes next.

- A grounded project manager fosters trust, which in turn strengthens your own belief:

We consistently emphasize that successful project management is rooted as much in who you *are* as in what you *do*. At the center of our philosophy is the idea of grounded leadership—being calm, self-aware, and values-driven. When a project manager is grounded, trust naturally follows, and that trust becomes a powerful source of confidence not only for the team, but for the leader as well. A grounded project manager brings stability to uncertainty. By staying present, listening carefully, and responding thoughtfully rather than reactively, leaders create an environment where people feel safe to speak honestly. We believe that trust is built through authenticity and consistency—when actions align with words, and decisions reflect shared values. Teams quickly sense when a leader is centered and reliable, even under pressure.

This trust transforms how teams engage. When people believe in their leader, they are more willing to collaborate, take responsibility, and navigate challenges together. Open communication replaces defensiveness, and problems surface earlier—when they are easier to solve. Trust, in this sense, becomes a performance enabler, not just a feel-good concept.

What is often overlooked is how this trust feeds back into the project manager. As team members respond with openness and commitment, the leader's own belief strengthens. Confidence grows—not from ego, but from evidence. Seeing the team respond positively reinforces the leader's sense of purpose, competence, and alignment. This creates a virtuous cycle: grounded leadership builds trust, trust fuels belief, and belief deepens grounding. Project management is ultimately about relationships. Tools and processes matter, but it is the grounded presence of the leader that sets the tone. By fostering trust through authenticity and calm leadership, project managers not only elevate their teams—they strengthen their own inner conviction, enabling them to lead with clarity, courage, and resilience.

• VISUALIZING SUCCESS

Success visualization entails intentionally using mental images of yourself achieving your goals as a source of motivation, guidance, and neural programming to make those goals a reality. Research shows that visualization practices can have physiological impacts, like reducing anxiety and fear responses to the possibility of unknown outcomes. We suggest some activities to help you visualize success:

- Before major meetings, presentations, or project launches, visualize yourself handling them effectively:

  We argue that effective project leadership begins long before a meeting starts or a project is launched. One of the most practical yet often underestimated disciplines they encourage is visualization, mentally rehearsing success before moments that truly matter. Before major meetings, presentations, or project kickoffs, grounded project managers take time to visualize themselves performing effectively. This is not wishful thinking; it is intentional preparation. By mentally walking through the event—entering the room calm and focused, communicating clearly, responding thoughtfully to challenges—the leader aligns mindset, emotions, and behavior. We emphasize that what happens internally often shapes external results.

  Visualization strengthens presence under pressure. When leaders have already "experienced" success mentally, they are less likely to be thrown off by unexpected questions or tense discussions. The mind recognizes the situation as familiar, reducing stress and increasing confidence. This inner calm translates outward, helping teams feel reassured and supported. Just as importantly, visualization reinforces belief. Seeing yourself lead effectively builds self-trust, which is essential in complex project environments. We think belief is not arrogance, it is grounded confidence rooted in preparation and self-awareness. When leaders trust themselves, others are more inclined to trust them as well. Over time, this practice becomes a leadership habit. Visualizing success before key moments encourages clarity of purpose, emotional balance, and intentional action. For project managers, it is a simple yet powerful way to lead with authenticity and resilience, showing up prepared not only in plan, but in presence.

- Mental rehearsal is a proven technique used by athletes and leaders to build inner belief:

  We draw parallels between effective project leadership and peak performance in sports. One practice that bridges both worlds is mental rehearsal—the deliberate visualization of successful performance. Used for decades by elite athletes and increasingly by leaders, mental rehearsal is a proven way to build inner belief and perform with confidence under pressure. Mental rehearsal works by preparing the mind before action is required. By vividly imagining key moments—communicating a clear message, responding calmly to resistance, guiding a team through uncertainty—leaders condition themselves for success. Englund and Bucero emphasize that belief grows through experience, and mental rehearsal creates a form of experience before reality unfolds. The mind learns what "success" feels like.

  This practice strengthens emotional control. When challenging situations arise, leaders who have mentally rehearsed them are less likely to react impulsively. Instead, they respond with clarity and composure. Just as athletes visualize executing the perfect move, project managers can rehearse difficult conversations, decision points, and leadership behaviors, reduce anxiety, and increase focus. Mental rehearsal also reinforces self-trust. Each intentional visualization sends a message to the leader's subconscious: *I am prepared.* Over time, this builds a quiet confidence grounded in readiness rather than bravado. Remember that inner belief is contagious—when leaders believe in themselves, teams sense it and respond with greater trust and engagement.

  In complex project environments, technical skills alone are not enough. Mental rehearsal equips leaders with the inner strength to match external demands. By adopting this proven technique, project managers cultivate belief, resilience, and presence—key ingredients for sustained leadership success.

• TAKING OWNERSHIP OF YOUR ROLE

Taking ownership of your role means being proactive, accountable, and contributing to the success of your organization. Shift from a follower role to a self-leader. Identify and apply the essence of who you are in mindset, talents,

skills, and experiences. Operate more from your internal beliefs and grasp of and closeness to reality rather than from external forces or voices from above. Believe that your internal thoughts are real and not arbitrary.
Here are some key points to consider:

- Understand your role: Recognize the purpose of your job and how it aligns with larger company goals.
- Accountability: Hold yourself responsible for your work and decisions and take initiative to resolve issues.
- Collaboration: Foster positive relationships with colleagues and communicate openly to ensure everyone is on the same page.
- Career Growth: Continuously seek opportunities to expand your skills and take on additional responsibilities.
- Enhance Job Satisfaction: Feel a sense of pride and fulfillment in your contributions to the organization, which can lead to greater job satisfaction.

• SURROUNDING YOURSELF WITH POSITIVE INFLUENCES

To surround yourself with positive influences, consider the following strategies:

- Build a supportive network: Focus on individuals who uplift and inspire you, including friends, family, and colleagues. Projects don't fail because of plans. They fail because of people disconnects. That's why one of the most critical skills of a successful project manager is the ability to build a supportive network. A supportive network starts with trust, not authority. Project managers who invest time in listening, keeping commitments, and showing genuine interest in people create relationships that hold when pressure rises. Strong networks are built early. Identifying a few key allies—a committed sponsor, a trusted technical expert, and an informal influencer—creates stability long before risks appear. These relationships are not transactional; they are nurtured through respect and consistency.

- Identify your influences: Reflect on who you turn to for positivity and motivation; write down their names to foster deeper connections.
- Evaluate your environment: Assess your current relationships and the media you consume.
- Identify negative influences and minimize their impact.
- Create a positive environment: Surround yourself with people who encourage and support your goals and engage in activities that foster growth and learning.
- Practice vulnerability: Share your goals and fears with supportive individuals to create deeper connections and encourage mutual growth.

PRACTICING GRATITUDE AND POSITIVITY

Gratitude makes a difference in a great leader. Gratitude takes a moment to reflect on how lucky you are when something good happens, whether it's small or big. You can use lots of words to describe feelings of gratitude: thankful, fortunate, humbled, appreciative, and blessed. I (Bucero) feel thankful to be a project manager, helping people and organizations to convert their dreams into successful projects. I (Englund) am grateful for all colleagues who provided feedback and opportunities to improve.

Here we share with you some good practices about gratitude:

- Start or end your day by listing three things you're grateful for or proud of.
- Express appreciation to your team and stakeholders.
- Focusing on what's working builds a mindset of abundance and resilience.

SEEKING AND GIVING MEANINGFUL FEEDBACK

We, as project professionals, receive feedback from time to time. But can we process it and use it in a positive way? Not always. Processing feedback goes far beyond listening to it in the moment and implementing it; it implies continuous reflection, dialogue, making some mistakes, and practice. While much of this happens because of the way we receive feedback, there is much we can do, too, to make sure we're processing feedback the right way. We have compiled some good practices on feedback:

- Ask for constructive feedback and use it to grow.
- Giving honesty, encouraging feedback to others boosts your self-awareness and leadership impact.
- Suspend a sense of certainty and be open to new insights.

ACTING WITH COURAGE

Courage inspires hope. In times of crisis and change, team members are most inspired by leaders who take initiative and risks. Sir Winston Churchill once said, "Success is not final, failure is not fatal; it is the courage to continue that counts." Courage, strength of character, and commitment are essential leadership traits in the project management field (and in most all others). As project professionals, we need to believe in our projects and exude enthusiasm and passion to team members and stakeholders. "A courageous leader always claims the power to initiate, act and risk, for courage means to act with sustained initiative," says Peter Koestenbaum, PhD, the Koestenbaum Institute, a leadership consultancy in Stockholm, Sweden.

As projects in Europe grow increasingly global and multicultural in scope, team members typically hail from all over the world. Project leaders need to be open and flexible in approaching other people, and cope with situations and people disparate from their own backgrounds. Be willing to reexamine and alter personal attitudes and perceptions.

Here are some thoughts on how to rev up your courage and build a better relationship with your team no matter where they are located:

Manage pressure. People are like melons, you never know what's inside until they're cracked apart. During difficult times, you discover what you truly believe. Regardless of what happens, though, keep your composure. Lead by example and remember your team is observing your every move. Some years ago, I (Bucero) managed a project in the north of Spain, about 400 kilometers away from home. I was on the customer site with my team all week and sometimes on the weekend. It felt like the project would never end, and I was distressed because my father was ill and died during the project life cycle. I was managing a strategic customer project, though, so I forced myself to encourage my team. Humor is a wonderful tool in some situations.

Balance hope with hard work. The more hope you have, the more work you'll put in to get what you want. But if you hope more than you do, you and your team members are likely to be disappointed, and your credibility is likely to suffer. If all else fails, switch gears. Changing strategy or aiming for an alternative target is not defeatism. If a project leader persists in a strategy that doesn't work, project team members may grow frustrated. Finding a new path of action or deciding on a different destination is a far more courageous move than continuing a problematic path.

I (Bucero) know colleagues managing project management office (PMO) implementation projects at multinational companies in Europe. In most of them, the PMO model was imposed by the organization's headquarters without ever considering the varying maturity levels from country to country, the different cultures and the range of stakeholder need. To keep their credibility among the team and to get the project finished successfully, many of these colleagues had to adopt different strategies and approaches as they implemented their PMOs.

Enjoy yourself. Sometimes the best answer is laughter. Courageous project leaders allow team members to have some fun, even when times are tough. Spanish people, for example, enjoy telling stories and jokes. Project managers who take things too seriously are much more likely to dig themselves deeper into a hole of despair and bury the entire team. So, go ahead, show some courage.

Here we summarize some best practices about courage.

- We build a belief when we do the right thing—even when it's difficult.
- Say no when necessary, raise complex topics, and challenge assumptions.
- Every act of courage strengthens your inner foundation.
- Know that fear may be present but push on regardless.

USING A MANTRA OR AFFIRMATION

- Many successful project managers use daily affirmations. Examples:
  - *"Today is a great day."*
  - *"I lead with integrity and clarity."*
  - *"I trust myself and my team to find solutions."*
  - *"I am aligned with abundance."*
- Say it daily, especially during challenging phases.
  What is your mantra? Please reflect upon it and gain more affirmation as a project manager.

Developing your personal development plan to build or reinforce your beliefs as a project manager takes persistence and patience. Use your passion to achieve it because you are an excellent project manager.

Tool
Here is an example of a short **Personal Development Plan** to help build belief as a project manager. It's structured over four weeks and can be repeated or adapted based on your needs.

BELIEF DEVELOPMENT PLAN FOR PROJECT MANAGERS

Duration: Four weeks.
Time commitment: 15–30 min/day.
Goal: Strengthen your self-belief, leadership presence, and resilience in project management.

◆ WEEK 1: CLARITY AND PURPOSE

Objective: Define your motivation and reconnect with your "why."

- **Daily (5 min):** Write your answer to:
  - *Why did I choose to lead projects (or engage in this endeavor)?*
  - *What value do I bring to my team?*
- **One-time:** Write a personal purpose statement (3–5 lines).
- **Reflection:** Identify one project success where you made a difference—write down how it made you feel.

End-of-week action: Share your purpose statement with a trusted colleague or mentor and ask for feedback.

◆ WEEK 2: CONFIDENCE THROUGH ACTION

Objective: Build belief by taking initiative and learning from action.

- **Daily (5–10 min):**
  - Reflect: *What did I do today that I'm proud of?*
  - Choose one small action to take ownership of tomorrow (e.g., initiate a difficult conversation, suggest an improvement).
- **Mid-week:** Ask for feedback from one team member: *What's one strength you see in me as a leader?*
- **End-of-week:** Choose a small risk to take (e.g., present a bold idea, challenge a status quo, delegate more).

End-of-week action: Write down what you learned and how you felt after taking that risk.

◆ WEEK 3: EMOTIONAL RESILIENCE AND GRATITUDE

Objective: Strengthen emotional self-awareness and positive mindset.

- **Daily (10 min):**
  - Write down three things you're grateful for.
  - Reflect on a challenge: *How did I respond emotionally? What could I do differently next time?*
- **One-time:** Identify your most significant current stressor and create a small coping plan (e.g., breathing, walking, talking to a peer, playing with a pet).
- **Mid-week:** Practice active listening in one meeting (no interruptions, just presence).

End-of-week action: Write a gratitude note to one team member for their impact on the project.

◆ WEEK 4: VISIBILITY AND AFFIRMATION

Objective: Strengthen your leadership voice and affirm your belief.

- **Daily (5 min):** Say your affirmation aloud (or write it):
  I am a capable, courageous leader who inspires others to succeed.
  (You can adjust this to fit your style.)
- **Mid-week:** Lead a team reflection where each person shares a recent win or learning.
- **One-time:** Create a "Belief Portfolio" (slide or page with photos, quotes, and reminders of your value).

End-of-week action: Write a short paragraph titled:

"Who am I as a project leader?"
We suggest repeating this plan quarterly or adapting the weekly focus based on project phases. Additionally, consider combining this plan with leadership books or coaching, if available, and sharing learnings with a peer or community of practice to reinforce your growth. In the next section is a proposed template to prepare a detailed belief development plan as a project manager.

Tools

BELIEF DEVELOPMENT PLAN FOR PROJECT MANAGERS

Duration: Four weeks.
Time Commitment: 15–30 min/day.
Goal: Strengthen your self-belief, leadership presence, and resilience in project management.

WEEK 1: CLARITY AND PURPOSE

Objective: Define your motivation and reconnect with your "why."

- Daily (5 min): Write your answer to:
  Why did I choose to lead projects?
  What value do I bring to my team?

- One-time: Write a personal purpose statement (3–5 lines).
- Reflection: Identify one project success where you made a difference—write down how it made you feel.
- End-of-week action: Share your purpose statement with a trusted colleague or mentor and ask for feedback.

WEEK 2: DEVELOP A VISION STATEMENT

Objective: Define a vivid description of a desired future state once the project is successful. It is unique to that project, and stakeholders will know when that new state is realized.

- Vision starts within. Vision comes from inside. Draw on your natural gifts and desires. Know yourself better; ask other people for feedback to get a fuller picture of your competencies and areas for improvement.
- Vision draws on your history. Vision grows from a leader's past and the experiences of the people around them.
- Vision meets others' needs. True vision is far-reaching, serving as an energy field for action.
- Vision helps you gather resources. The greater the vision, the more winners it has the potential to attract.
- Know where vision comes from. To find a vision that is indispensable to leadership, become a good listener. Listen to several voices:
  - The inner voice: Do you know your life's mission as a project manager?
  - The unhappy voice: Discontent with the status quo is a great catalyst for vision.
  - The successful voice: Nobody can accomplish great things alone. Fulfilling a big vision requires a good team. But you also need advice from someone who is ahead of you in the leadership journey. If you want to lead others to greatness, find a mentor. Do you have an advisor who can help sharpen your vision?
  - The higher voice: Although it is true that your vision needs to come from within, do not let it be confined by limited capabilities.
- End-of-week action: Share your vision statement with a trusted colleague or mentor and ask for feedback. Use a "Wow!" response as an appropriate metric.

WEEK 3: EMOTIONAL RESILIENCE AND GRATITUDE

Objective: Strengthen emotional self-awareness and positive mindset.

- Daily (10 min):
  Write down three things you're grateful for.
  Reflect on a challenge: How did I respond emotionally? What could I do differently next time?
- One time: Identify your most significant current stressor and create a small coping plan.
- Mid-week: Practice active listening in one meeting.
- End-of-week action: Write a gratitude note to one team member for their impact on the project.

WEEK 4: VISIBILITY AND AFFIRMATION

Objective: Strengthen your leadership voice and affirm your belief.

- Daily (5 min): Say your affirmation aloud or write it:
  "I am a capable, courageous leader who inspires others to succeed."
- Mid-week: Lead a team reflection where each person shares a recent win or learning.
- One-time: Create a "Belief Portfolio" (slide or page with photos, quotes, and reminders of your value).
- End-of-week action: Write a short paragraph titled: "Who I am as a project leader."

Ongoing Maintenance Tips

- Repeat this plan quarterly or adapt a weekly focus based on project phases.
- Combine this plan with leadership books or coaching if available.
- Share learnings with a peer or community of practice to reinforce your growth.

STRATEGIES TO STRENGTHEN YOUR BELIEFS

Developing your beliefs as a project manager is a continuous process that involves reflection, learning, adopting, adapting, and applying based on your experiences. Here are a few strategies to help develop and strengthen your beliefs. See Figure 5.1.

1. Self-Reflection: Take time regularly to reflect on your experiences.
   - What has worked well in your projects?
   - What could be improved?

   Self-reflection helps you understand what you value most in your leadership style and project management approach. Self-reflection is key to evolving your beliefs over time.
2. Seek Feedback: Solicit feedback from your team and stakeholders. Feedback can help you understand how your approach impacts others and guide you in shaping your beliefs. Knowing how your team perceives your leadership can offer insights into areas of strength and areas for growth. For example, running Retrospective Analysis at the end of each project phase is helpful for analyzing: What did you do well? What did you not do well? And what did you learn?
3. Learn from Others: Learn from other experienced project managers and leaders. May learn through books, workshops, seminars, or mentoring relationships. Listening to their experiences and perspectives can challenge and expand your own beliefs. One effective practice, in my experience (Bucero), is to organize periodic symposia among project managers within the organization and invite executives to attend. I (Englund) like to observe how other leaders handled challenging situations or conversations.
4. Stay Open to New Ideas: Embrace a mindset of continuous improvement. As the project management field evolves, so should your beliefs. Be open to new methodologies, tools, and techniques that can enhance your effectiveness and broaden your perspective. We embrace a "you could be my teacher" culture, which involves observing every project stakeholder and learning from them.
5. Adopt a Growth Mindset: Approach challenges as opportunities for growth. When things don't go as planned, see them as learning

FIGURE 5.1
Strategies to strengthen your beliefs.

experiences that can shape and refine your leadership beliefs. For example, some good practices for us (Englund & Bucero) include encouraging project managers to submit project experiences to International Congresses, creating PM Initiatives, hosting PM Forums, and participating in Zoom sessions.

6. Practice Empathy: Develop a deep understanding of your team members, their strengths, and challenges. Empathy helps you refine your belief in people and adjust your leadership style to support their success. Understand where people come from. For example, we recommend spending time with project team members to get to know them better and create a healthy project atmosphere. What experiences contributed to a current mindset?
7. Lead by Example: Show through your actions how you value your beliefs. For example, if you believe in Teamwork, foster collaboration by encouraging open communication and giving team members opportunities to lead. For example, if you feel a positive attitude can impact project success, strive to maintain a positive outlook every day, even in challenging circumstances.
8. Establish Core Values: Identify the values that are non-negotiable for you, such as integrity, trust, and accountability. Ensure your actions and decisions consistently reflect your values. For example, based on our experiences, we establish authenticity (say what you believe) and integrity (act on what you say) as core values in every project we manage. By being intentional in these areas, you can continue to develop your beliefs, which will guide how you lead your projects and your team. But what is the impact of the project manager's beliefs in their project context?

Impact of PM Beliefs

A project manager's beliefs significantly impact a team's inspiration, motivation, and overall performance. Your belief that projects succeed through Teamwork and your focus on positivity create a strong foundation for inspiring your team.

Here's how a project manager's beliefs may influence team members:

1. Belief in Teamwork → Increased Collaboration
   When a project manager genuinely believes that Teamwork is the key to success, team members are more likely to value collaboration and support each other.
   Your habit of rotating leadership roles and encouraging feedback strengthens this belief, making team members feel empowered and responsible.
   Ensure that metrics support teamwork, not just individual performance.

2. Belief in a Positive Attitude → Higher Resilience
   If a project manager stays optimistic, even in difficult phases, the team is more likely to adopt a problem-solving mindset instead of feeling discouraged.
   Focusing on the 'whole part' of the bottle helps team members reframe challenges as opportunities.

3. Belief in Individual Potential → Increased Confidence
   When a manager trusts their team and gives them the benefit of the doubt, employees feel more confident in their abilities.
   Your strategy of letting team members lead meetings and present their roles reinforces their sense of ownership.

4. Belief in Gratitude → Stronger Team Bonds
   Encouraging team members to express appreciation fosters a positive team culture.
   When people feel valued, they are more engaged and willing to go the extra mile.

Final Impact: A Motivated and Inspired Team and Your Leadership Style, rooted in these beliefs, creates a work environment where people feel supported, appreciated, and motivated to contribute. Even though not all team members react positively to challenges, your continuous encouragement helps shift their mindset over time.

Noted author Greg Githens wrote quite a bit about beliefs in his book *Strategic Thinking:*

I suspect I'm defining them differently than you. For me, beliefs are quite malleable and change with the evidence and data of the specific situation. Based on what I believe to be reality and my interests, I take appropriate/reasonable action. My work on strategic thinking suggests that many people develop a false confidence and take unreasonable actions because they don't search very far for disconfirming evidence and sometimes see skepticism as a weakness. And they are mentally lazy, too.

I take it that you are using belief in the sense of confidence in one's self or in the sense of a mental model. When one has a "mindset," they are often inflexible; that's why I like the alternative concept of "worldview." In my book, I developed the idea of "Perspective" and defined it as "personality plus point of view." Which is to suggest that some aspects of our conduct are temperament that are mostly baked-into our personalities and some aspects of our conduct are filtered by the way that we select data and make sense of it.

Tool

Belief Development Template

Below is a proposed template designed to help you, as a project manager, develop and articulate your core beliefs, values, and guiding principles. This template is structured to facilitate self-reflection and alignment with your professional practice.

■ Project Manager Belief Development Template

Name: ____________________________________

Date: _____________________________________

Personal Core Values

Identify and list the fundamental values that guide your behavior and decisions as a project manager.

- -- ------------------------------------
- -- ------------------------------------
- -- ------------------------------------

Professional Beliefs

Articulate the beliefs you hold about effective project management, leadership, and team collaboration.

- --
- --
- --

Alignment Assessment

Evaluate how your personal values align with your professional beliefs and practices.

- Are there any discrepancies between your values and your actions? --
- How can you ensure consistency between your beliefs and your professional conduct? --

Belief Development Goals

Define specific goals to develop or reinforce beliefs that will enhance your effectiveness as a project manager.

- Goal 1: ____
- Goal 2: ____
- Goal 3: ____

Action Plan

Outline actionable steps to achieve your belief development goals.

- Action Step 1: ____
- Action Step 2: ____
- Action Step 3: ____

Progress Review
Schedule regular intervals to review your progress and make necessary adjustments.

- Review Date: ____
- Observations: ____

This template is inspired by best practices in personal and professional development, emphasizing the importance of aligning one's values and beliefs with their professional role.

6

Keeping Alive Your Belief

Men willingly believe what they wish.

Julius Caesar (100 BC–44 BC)

I (Bucero) am aware that some project managers mistakenly believe that all feedback is essentially negative and confuse it with criticism. When we believe it is necessary, we provide both good and negative comments. As project managers, it is our responsibility to get the most out of our employees. As leaders, it is our responsibility to help our team members achieve outstanding project outcomes. Why not use feedback to keep your project stakeholders communicating in a clear, succinct, and fluid manner?

Every team member needs your feedback, but also you need some feedback from them. We think it is not possible to see one's own neck without assistance; therefore, obtaining additional information and alternative perspectives from others can be valuable when assessing one's beliefs or assumptions. In my professional career I (Bucero) always ran feedback sessions every week with my team members and with my project sponsor. At the beginning I felt so nervous and intimidated. I listened to some things I did not like. However, it took a series of feedback sessions where I learned a lot, and that helped me move forward as a project professional and learn from my mistakes. And in fact, I always make some mistakes managing projects and programs, but I always try to learn from them.

We found some executives that never asked for feedback. Sentences like "*No thanks, I don't need your opinion, I have been involved in many projects for many years, you cannot teach me anything.*" They are still there in the same position and making the same mistakes; they never listen to their people so they will not learn. Everybody can be our teacher is one of our preferred sentences and every day we can learn something from somebody. Don't wait

DOI: 10.1201/9781003730736-6

any more, get feedback sessions organized with your team members in your projects. People can give you positive and negative feedback. Be focused on learning from feedback that other people provide. We found some useful guidelines to giving positive feedback that we would like to share with you:

- **Be a cheerleader for your people**. Encourage them persistently. Find ways to encourage each other. Caring of your people is one of your obligations as a good leader.
- **Back people up and build people up**. Help people develop themselves professionally and personally.
- **Say thank you frequently**. Find the right occasions to do so.
- **Choose your words** to reinforce other persons' self-worth by borrowing their relating style words that are meaningful to them but not necessarily to you.
- **Give/praise** in private and in public. Use verbal and written praise.
- **Let your praise of your people** be heard by clients or customers and be sincere in giving that praise.
- **Avoid words/choices that include the word "but"** which negates everything that comes before.
- **Watch your timing** and emotional conditions of others
- At the beginning of my career I (Bucero) was so shy, and giving negative feedback was sometimes more difficult than giving positive feedback. However, based on our experiences, we also have some best practices for you that helped us managing people in projects:
  - Remember that most people want to do the right thing
  - Show appreciation for the part that was done well
  - Never personalize the criticism
  - Avoid the word "you" as much as possible
  - Share "I" statements as examples
  - Focus discussions on improvements needed or on solutions
  - Focus on the problem not on the people
  - Focus on learning from facts and situations
  - Provide outside information as a reference point wherever possible "These conditions are different because ..."
  - Use your patience
  - Use an appropriate style of communication according to people's personalities
  - Use your filters when listening to responses to negative feedback or criticism

RECEIVING FEEDBACK

To improve your leadership style and make sure that your values have a good influence on your team and projects, get input on your project management beliefs. We argue that if you want to maintain your project management principles, keep getting better. To build team trust, create a safe space for criticism. We present you with a useful image in Figure 6.1 that you can adhere to (note the curves and uncertainty about what lies beyond the curve).

Lead with Psychological Safety

People don't speak up when they fear blame, judgment, or consequences.

- **Normalize imperfection**: Remind your team that feedback—positive or critical—is expected and valued by all stakeholders.
- **Admit your own mistakes** openly to model vulnerability. If you made a mistake, we encourage you to recognize it face to face.
- **Encourage others by saying**: "Your honest feedback helps me grow just like I try to help you grow."

Make Feedback a Two-Way Habit

"Feedback shouldn't flow only top-down." Sometimes executives are not ready to receive feedback from their project managers.

FIGURE 6.1
Road map.

- Ask for feedback regularly, not just during retrospectives or at project closure.
- Use consistent prompts like: "What's one thing I could do differently to support you better?"
- Include structured feedback in:
  - One-on-ones
  - Sprint reviews
  - Weekly or monthly check-ins

Establish Ground Rules for Respectful Communication

"Feedback should be honest, but always kind."

- Co-create team guidelines:
  - Focus on behaviors, not personalities
  - Assume positive intent
  - Be specific, not vague
- Train the team on using **nonviolent communication** or the **SBI model** (Situation–Behavior–Impact).

Use Multiple Feedback Channels

"Some people are more comfortable writing than speaking." Some people are scared when they need to present.

- Provide anonymous feedback options (like a Google Form).
- Create a "Feedback Friday" Slack thread or a physical board.
- Offer private spaces like 1:1s for more sensitive comments.

Act on Feedback Promptly

"Feedback that disappears into a void kills trust."

- Acknowledge every piece of feedback with a thank you.
- Share what you're doing about it—even small steps.
- Example: "Last week someone shared that our meetings run too long. Starting today, I'll limit updates to 1 minute per person."

Celebrate Feedback Moments

"Appreciating the messenger builds a feedback-positive culture."

- Recognize team members who give constructive feedback.
- Publicly thank someone for helping you improve your leadership style.
- Reinforce that feedback is a gift, not a complaint.

Link Feedback to Growth and Shared Purpose

"Make it clear that feedback supports the project and the people."

- Share how feedback has influenced your beliefs or practices.
- Use team language:
  "We're all here to succeed together—your feedback helps me stay true to that belief."

EXTRA TIPS JUST FOR YOU

Based on your approach to leadership (positive attitude, teamwork, trust-building), consider also:

- Rotating feedback facilitators to give everyone a voice.
- Having your team reflect not only on what you could improve, but what values you consistently model well (e.g., optimism, gratitude, support).

Ask Specific, Thoughtful Questions

- Instead of general questions like "Do you think my approach works?," try:
- "How has my focus on motivation impacted your engagement in this project?"
- "Do you feel my leadership style helps the team work better together?"
- "Are there areas where my approach could be improved?"

We consider that asking specific, thoughtful questions is a powerful way to keep your beliefs as a project manager alive in everyday practice. Your beliefs—like valuing people, maintaining a positive attitude, fostering teamwork, and giving trust—can be nurtured by the *questions you ask* consistently. Here's how to do it with purpose:

Anchor Each Question in a Core Belief, Table 6.1

TABLE 6.1

Anchoring Beliefs

Belief	Example Question
Positive Attitude	"What's something positive that happened this week—even if the week was hard?"
People First	"Who supported you most this activity, and how can we acknowledge them?"
Trust and Empowerment	"Is there a decision you'd like to make more independently next time?"
Teamwork and Belonging	"Do you feel your role in the team is clearly understood and appreciated?"
Growth and Recognition	"What's one thing you learned or improved on recently that you're proud of?"

Use Questions That Trigger Reflection, Not Defense

Avoid vague or judgmental questions. Instead of "Why didn't this go well?," try:

- "What do you think we could try differently next time to avoid this?"
- "What support would have made a difference for you?"

This keeps the tone constructive and aligned with your belief in positive leadership.

Incorporate Questions into Regular Routines

Integrate your belief-driven questions into:

- Stand-ups ("What's something you're excited about this week?")
- Retrospectives ("Who went above and beyond for the team this sprint?")
- One-on-ones ("How can I support your personal growth?")
- Feedback Friday ("What's one thing I did this week that helped—or didn't help—you?")

Ask About Emotional and Cultural Climate

To protect psychological safety and motivation:

- "Do you feel heard and valued on this project?"

- "Are there any frustrations or blockers we haven't talked about?"
- "Is the way we're working aligned with our team values?"

Invite Feedback on Yourself, Thoughtfully

Model humility and growth by asking:

- "What's something I could do differently to better live up to the values I promote?"
- "Am I giving you enough space—or too much?"
- "Have you seen me act in a way that contradicts my beliefs? I'd like to know."

Follow Up on What You Hear

Ask follow-up questions that reinforce your care and curiosity:

- "You mentioned feeling disconnected—what small change could help you feel more included?"
- "Thanks for your honesty last time—have things improved since then?"

ANOTHER TIP: KEEP A PERSONAL "BELIEF-TO-QUESTION" LIST

Make a small table for yourself like this, and revisit it monthly Table 6.2:

Would you like a printable **Belief-to-Question Cheat Sheet** for your desk or team room?

Use Multiple Feedback Channels

- One-on-One Meetings: Some team members may feel more comfortable sharing their thoughts privately.
- Anonymous Surveys: Provide a way for people to give honest feedback without fear of judgment.
- Team Retrospectives: Facilitate discussions about what's working and what needs improvement.
- Peer Reviews: Ask other project managers or sponsors for their perspectives on your approach.

TABLE 6.2

Belief Questions

My Belief	My Go-To Question
People are the key	"How are you feeling about your contributions to the project?"
Feedback builds trust	"Is there anything I should stop, start, or continue doing?"
Gratitude boosts morale	"Who deserves a special thank-you this week?"

KEEP YOUR BELIEFS ALIVE

To keep your beliefs, as a project manager, alive using multiple feedback channels, you need to build a *living feedback ecosystem*—one where your team sees that your beliefs (trust, positivity, teamwork, growth, recognition) are *actively nurtured* and not just words in a kick-off slide. Here's how we suggest you do it:

One-on-One Conversations (Private, Personal Feedback)

Ideal for trust, vulnerability, and growth-focused feedback.

- *Ask belief-aligned questions like:*
  "What's one thing I could do better to support your work or well-being?"
- *Use this channel to:*
  - *Address sensitive issues*
  - *Follow up on feedback from other channels*
  - *Reassure team members of your commitment to personal growth*

Why it works*: People open more in safe, individual spaces—especially if they fear judgment in public forums.*

Anonymous Feedback Forms (Safe, Honest Insight)

Ideal for surfacing hidden concerns and measuring alignment with your beliefs.

- Use tools like Google Forms, Typeform, or Slido with prompts such as:
  - "Where do you see me living up to our shared values?"
  - "Where do you feel we are drifting from our project purpose or culture?"
- Do this quarterly or after critical project phases.

Why it works: Encourages honesty from more introverted or hesitant voices.

Team Retrospectives or Feedback Circles (Collective Learning)

Ideal for reinforcing beliefs around collaboration, mutual respect, and transparency.

- Dedicate 10–15 mins to belief-driven questions:
  - "What did we do in this sprint that showed trust in each other?"
  - "Where did we stray from our principle of open communication?"
- Use sticky notes, Miro, or digital retrospectives for group input.

Why it works: Builds shared responsibility and normalizes feedback as a team tool, not a top-down one.

Instant Messaging Channels (Lightweight, Ongoing Feedback)

Ideal for gratitude, recognition, and micro-feedback in the flow of work.

- Create a "#feedback" or "#shoutouts" channel on Slack or Teams.
- Prompt with questions like:
  - "Who made your job easier this week?"
  - "What small act of leadership did you notice in someone else?"

Why it works: Reinforces positivity, encourages peer recognition, and makes feedback continuous.

Feedback Rounds During Stand-Ups (Fast, Consistent Pulse)

Ideal for checking emotional climate and maintaining alignment.

- End the week with a quick round:
  - "One word to describe how you felt about this week."
  - "One thing you appreciated or struggled with."

Why it works: Keeps feedback lightweight and regular, avoiding emotional buildup.

Rotating Feedback Hosts or "Voice of the Team" Roles

Ideal for empowering others and reinforcing trust.

- Let different team members lead feedback sessions.

- Give one person a "voice of the team" role for a week or sprint, tasked with gathering and sharing anonymous thoughts.

Why it works: Shows that everyone's voice matters and feedback is everyone's responsibility.

Visible Feedback Boards (Physical or Digital)

Ideal for transparency, shared values, and celebrating alignment with beliefs.

- Use Kanban-style boards or Miro templates with columns like:
  - "What we're doing well"
  - "Where we can grow"
  - "What we appreciate"

Why it works: Makes feedback visual and participative, fostering a sense of team ownership.

THE KEY TO SUCCESS: ACT ON FEEDBACK PUBLICLY

Whatever channel you use:

- **Acknowledge the feedback** you receive.
- **Connect your actions to your beliefs.**

Someone mentioned they didn't feel recognized last sprint. Since people are the key to our success, I'll be more intentional about daily acknowledgments.

Observe Team Reactions and Behaviors

- Pay attention to non-verbal cues, engagement levels, and overall team morale.
- If your team seems disengaged despite your motivational efforts, it might indicate a need for adjustments.

Observing your team's reactions and behaviors is a crucial way to *validate and reinforce your beliefs* as a project manager. Your beliefs—like positivity, trust, teamwork, recognition, and growth—aren't just slogans. They're visible (or not) in *how people act, interact, and respond.* Here's how to watch closely and interpret what you see:

Use Observation as a Mirror of Your Beliefs

Ask yourself:

- **"Are my beliefs reflected in how the team behaves, even when I'm not around?"**

Example Beliefs → Observable Behaviors Table 6.3:

Read the Emotional Climate

Your team's emotions and energy are constant feedback loops. Observe:

- Facial expressions and tone in meetings (confused, disengaged, enthusiastic?)
- Participation levels (Are only a few people talking?)
- Peer-to-peer engagement (Are they talking directly with each other—or always through you?)
- Body language (Closed off? Defensive? Relaxed and open?)

Tip: Watch especially during retrospectives, stressful phases, or after giving tough feedback. These moments are windows into the team's real psychological safety.

TABLE 6.3

Belief Behaviors

Belief	Watch for This Behavior
Trust	Team members take initiative, ask for help without fear, admit mistakes without blame.
Positive Attitude	Team members look for solutions instead of assigning blame when problems arise.
Recognition Matters	People give shout-outs to each other without prompting.
Teamwork and Belonging	Members offer support proactively, no one is left isolated or unheard.
Growth is Encouraged	People seek feedback, share learning moments, and are open to stretch assignments.

Notice What's Said—and Unsaid

- Are your values echoed in team language?
  - Do people say things like: "Thanks for supporting me," or "Let's learn from this"?
  - Do people use "I" or "we" statements?
- Are team members silent in feedback sessions? That might signal fear, doubt, or misalignment with your belief in openness.

Look at Behavior Over Time, Not Just Moments

Short-term reactions are important—but patterns reveal the truth.
Try this:

- After each major meeting, ask yourself:
  - "Who looked engaged? Who was distracted or disconnected?"
  - "Did anyone change their usual behavior—speak less, hesitate, withdraw?"
- After a sprint or milestone, reflect:
  - "Are people collaborating more naturally now?"
  - "Are gratitude and ownership growing or fading?"

Document Informal Observations

One good practice we suggest is to keep a small leadership observation journal. Log things like:

- "Today, Ana helped Marcos without being asked—shows strong teamwork culture."
- "No one spoke in our last stand-up—might need to reenergize trust and engagement."

This helps you make data-informed adjustments to your leadership behavior.

Create Safe Experiments and Observe Reactions

Example:

- Try handing over leadership of a planning meeting to a team member.

- Watch how the team responds: Are they supportive, hesitant, empowered?

These small experiments test how alive your beliefs really are in team culture.

Close the Loop with Reflection Questions

After your observations, bring your team into the reflection:

- "I've noticed we've been quieter lately—how are you all feeling?"
- "I saw some great peer support this week. How can we keep that going?"

You connect your beliefs with what you observe, which shows leadership that listens and adapts.

BOTTOM LINE

Beliefs stay alive not because you say them—but because your team starts to live them.

Observation is how you spot that quiet truth.

Encourage Constructive Criticism

- Normalize feedback as a tool for growth rather than a personal critique.
- When someone challenges your belief, ask them to explain their perspective and suggest alternatives.

Encouraging constructive criticism is one of the best ways to keep your beliefs as a project manager alive—especially if your values include **trust, transparency, growth, and team empowerment**. When people feel safe to challenge ideas and give honest input, they become co-owners of your project culture, not just passengers. Here's how to do it effectively:

Start with Your Own Example (Model It First)

People won't offer constructive criticism unless they see that *you* are open to it.

- Ask regularly:

"What's something I did last week that didn't help the team as much as I thought it would?"

"Was there a better way I could have handled that issue?"

- Thank people openly when they give you feedback.
  "Thanks for the honesty—that helps me grow."

This proves you practice what you believe: openness, humility, and growth.

Create Psychological Safety

- Make it clear that mistakes and feedback are learning opportunities.
- Separate people from the problem:
  "Let's challenge the process, not each other."
- When criticism arises, never punish it—reward it by listening, acting, and giving credit.

This supports your belief in trust and transparency.

Use Constructive Prompts to Invite Criticism

Avoid: "Any feedback?" → too vague, too risky.
Use:

- *"What's one thing we could improve as a team?"*
- *"What didn't feel right in that last sprint?"*
- *"Where did I miss the mark this week?"*

Also try structured formats like:

- Start—Stop—Continue
- Rose—Thorn—Bud
- One Improvement, One Appreciation

These tools reinforce your belief in continuous improvement.

Create Feedback Moments—Don't Wait for Them

- Dedicate time in retrospectives to "What could have gone better?"
- In 1:1s, always ask:

"How are we doing as a team?"
"What's not working for you?"

This shows that feedback is *expected*, not exceptional.

Train the Team in Giving and Receiving Feedback

Host a short session on:

- How to give feedback using "I" statements, facts, and suggestions.
- How to receive feedback with curiosity, not defensiveness.

For example:
"You never listen in meetings."

Rather, "I felt unheard when I shared my idea and we moved on without discussion."

This keeps alive your belief in respectful communication and growth.

Encourage Peer-to-Peer Criticism, Not Just Top-Down

Make peer feedback part of your culture:

- Use "Appreciation + Suggestion" during sprint reviews.
- Assign **feedback buddies** for informal feedback rounds.

One thing I admire about your work this sprint, and one thing I think you could try differently.

This strengthens your belief that everyone is a leader in their own way.

Use Feedback as Evidence of Your Beliefs in Action

When someone offers constructive criticism:

- Celebrate it publicly.
- Reflect it back to the team:

> "Cristina pointed out something important about how we make decisions. That honesty shows we're living our belief in openness."

You reinforce that your beliefs are not just theoretical—they shape behavior.

GOOD PRACTICE: MAKE IT A RITUAL

- Add a "Feedback Reflection" section to your team meetings.
- Include "Best Piece of Feedback This Month" in your newsletter or Slack.
- Rotate a "Feedback Champion" role to normalize it.

Reflect on Results and Adjust

- Look at project success rates, team satisfaction, and stakeholder engagement to see if your approach is yielding the right outcomes.
- If certain aspects aren't working, adapt while staying true to your core belief that people drive project success.

Reflecting on results and adjusting your approach is how you breathe life into your project management beliefs every day. Your core beliefs—like trust, positive attitude, team empowerment, and continuous improvement—will only thrive if you consistently align your actions with outcomes and make space to learn. We suggest a practical guide to doing just that:

Pause and Review: What Happened?

"Results don't lie—but they don't speak unless you ask the right questions." After each milestone, sprint, or significant event, ask:

- What did we accomplish?
- What worked better than expected?
- What didn't go according to plan?

This step shows your belief in self-awareness and truth-seeking.

Reflect Through the Lens of Your Beliefs

Use your beliefs as a filter to interpret the results Table 6.4.
This is where beliefs move from *abstract values* to *daily decisions.*

Ask for Team Input (Collaborative Reflection)

To reflect alone is wise. To reflect together is powerful.

In retrospectives or feedback sessions, ask:

- "Where do you feel we lived up to our project values?"
- "What moments felt misaligned with who we say we are?"
- "What small changes would make a big difference next time?"

This shows your belief in team ownership and learning from each other.

Adjust Behavior and Systems

Reflection is useless without course correction.

- If positivity dropped, introduce a gratitude ritual.
- If trust was low, move some decisions closer to the team.
- If recognition faded, build a shout-out section into meetings.

These small adjustments reinforce that beliefs guide behavior.

TABLE 6.4

Belief Questions

Belief	Reflection Question
Trust matters	Did the team feel safe to speak honestly? Did we empower people to make decisions?
Positive attitude wins	Did we stay resilient during setbacks? Did we focus on solutions instead of blame?
Teamwork is key	Did we collaborate effectively? Did we support each other under pressure?
Recognition drives people	Did we celebrate progress? Did everyone feel seen and appreciated?
Growth mindset	What did we learn? Did we challenge ourselves and try something new?

Track Changes and Celebrate Progress

Create a simple "Belief Alignment Tracker:"

Sprint	Trust Score (1–5)	Team Feedback on Collaboration	Actions Taken	Result

Use it monthly to:

- Monitor the health of your team's culture
- See which beliefs are thriving and which need care
- Keep your values visible and measurable

This reflects your belief in accountability and continuous growth.

Make Reflection a Habit, Not a Reaction

Turn reflection into a recurring rhythm:

- Personal journaling: 10 mins weekly—What did I learn? Did I lead with my values?
- Team check-ins: End each sprint with 3 prompts:
  1. What did we achieve?
  2. How did we behave?
  3. What will we do differently?

Habits like these keep your beliefs *present and active.*

FINAL THOUGHT

You don't keep your beliefs alive by stating them—you keep them alive by aligning them with reality, every week.

Show Appreciation for Feedback

- Acknowledge and thank people for their input, even if you don't agree with everything they say.
- Demonstrating that you value feedback will encourage more honest discussions in the future.

Showing appreciation for feedback is essential to keeping your project management beliefs alive, especially if you value trust, transparency, growth, and psychological safety. When team members see that feedback is welcomed and *appreciated*, not just tolerated, they'll feel more confident and engaged—which reinforces the environment you're working to build.

Here's how to show appreciation genuinely and effectively:

Say "Thank You" Right Away—And Mean It

When someone gives you feedback (positive or critical):

- Acknowledge it immediately.
- Use grateful and specific language:

  "Thank you for being honest about that—it shows you care about the team."
  "I appreciate your courage in bringing that up. That's exactly what helps us grow."

This shows that your belief in openness and learning isn't just talk—it's a lived value.

Follow Up Thoughtfully

Let people know what you did with the feedback:

- "You mentioned our stand-ups felt rushed; starting tomorrow, we'll try a new time limit structure."
- "Thanks to your input, I changed how we frame sprint goals."

This reinforces your belief that everyone's voice matters, and that feedback drives improvement.

Publicly Recognize Feedback Givers

In team meetings, Slack, or retrospectives:

- Highlight moments when feedback made a difference.
  "Shout out to José for his great insight on how we handle testing cycles—it helped us improve our process."

This aligns with your belief in recognition and team contribution.

Tip: Recognize both *what* was said and the *courage* it took to say it.

Use Personal Notes or Quick Messages

Send a quick private message or handwritten note:

- "Your feedback helped me reflect more deeply than you know. Thank you."
- "It takes guts to say what you said today. I admire that."

These build trust and strengthen person-to-person connection, which reinforces your belief in empathy and people-first leadership.

Act on Feedback—That's the Ultimate Form of Appreciation

Words matter, but action is everything.

- Change what needs to change.
- Test suggestions where appropriate.
- Credit the person when results improve.

"We now do daily recaps because of feedback from Clara. It's been a game changer."

Living your beliefs through action shows that feedback is **a gift, not a threat**.

Make Feedback Appreciation a Ritual

Build habits that show continuous gratitude:

- End retros with:
  "One piece of feedback I'm grateful for this sprint."
- Add a "Thank You Wall" in your office or Slack channel.
- Rotate a "Feedback Appreciator" role in retrospectives.

These consistent acts make appreciation part of your team's culture.

Connect Appreciation to Beliefs Out Loud

Be intentional:

"When you gave that feedback, it reminded me why I believe in psychological safety so deeply. That's how we improve as a team."

This ties feedback to your *values*—and keeps them visible, alive, and active in your leadership.

BONUS: APPRECIATION PHRASES YOU CAN USE

- "That feedback gave me a new perspective—thank you."
- "I'm grateful you cared enough to speak up."
- "You helped me grow as a leader. I value that."
- "Thanks for helping the whole team level up."

Bottom Line:

APPRECIATION TRANSFORMS FEEDBACK INTO CULTURE

It keeps your beliefs not just *stated* but *shared.*

By actively seeking and embracing feedback, you ensure that your project management beliefs remain relevant, effective, and continuously evolving to support your team's success.

SUSTAINING AND MAINTAINING YOUR BELIEFS

Sustaining and maintaining your beliefs, especially in a professional and leadership context, requires a combination of self-awareness, continuous reinforcement, and adaptability. Here are some strategies to help you stay committed to your beliefs:

1. Stay True to Your Core Values
    - Define and document your core beliefs. In your case, inspiring, motivating, and leading people to project success is central.

- Regularly reflect on why these beliefs matter to you and how they align with your personal and professional mission.

2. Practice What You Preach
   - Lead by example—your team will follow your attitude and actions more than your words.
   - Stay consistent in how you apply your beliefs, whether it's through motivation, stakeholder management, or fostering teamwork.
3. Reinforce Your Beliefs Daily
   - Use affirmations or daily reflections to remind yourself of what you stand for.
   - Keep a gratitude practice, as you already encourage your team to reflect on positive achievements each day.
4. Surround Yourself with Like-Minded People
   - Engage with colleagues, mentors, or communities that share or support your values.
   - Attend events, read books, and follow thought leaders who align with your perspective.
5. Adapt Without Compromising Core Values
   - While being steadfast is important, flexibility helps sustain your beliefs in changing circumstances.
   - Adapt your motivational approach based on team needs, just as you do now.
6. Educate and Evolve
   - Continue learning about leadership, motivation, and project sponsorship.
   - Stay informed about industry trends and best practices to refine your approach without losing sight of your core beliefs.
7. Measure the Impact of Your Beliefs
   - Observe how your leadership style influences project success and team morale.
   - Gather feedback from your team to assess whether your beliefs are driving positive results.
8. Overcome Challenges with Resilience
   - Not everyone will immediately align with your mindset—stay patient and persistent.
   - Use setbacks as learning experiences to reinforce your commitment rather than weaken it.

You've already built strong habits to sustain your beliefs, such as team reflections, giving feedback, and maintaining a positive outlook. The key is to keep evolving while staying true to your mission of inspiring people for project success.

INFLUENCE OF PM BELIEFS

A project manager's beliefs strongly influence their ability to gain **executive support**, which is crucial for project success. Here's how different beliefs impact executive engagement:

1. Belief in the Strategic Value of Projects
   - If a project manager believes their project contributes to the organization's strategy, they will align their messaging with executive priorities.
   - This helps in securing sponsorship, funding, and leadership backing.
2. Belief in the Power of Influence Over Authority
   - Executives respond better to influence rather than demands. A project manager who believes in influencing through value rather than formal authority can gain stronger support.
   - By showing business impact and using data-driven arguments, they engage executives more effectively.
3. Belief in Communication and Relationship Building
   - Project managers who believe in active engagement with executives will regularly update them on progress, risks, and key decisions.
   - This transparency builds trust and ensures continued executive commitment.
4. Belief in Sponsorship as a Key Success Factor
   - A project manager who understands the importance of sponsorship will proactively seek executive involvement rather than assuming support will come naturally.
   - They educate executives on their role in project success, making sponsorship more active and effective.
5. Belief in Solution-Oriented Thinking
   - If a project manager believes that problems are opportunities for solutions, they will approach executives with proposals rather than just issues.

    - This increases executive confidence in the project and encourages their involvement.
6. Belief in Adaptability and Business Agility
    - Executives prioritize business agility and results. A project manager who embraces change and innovation will gain stronger backing from leadership.
    - They will frame project decisions in terms of business impact, making it easier to get executive buy-in.

KEY THOUGHTS

A project manager's mindset shapes executive engagement. When they position themselves as strategic partners rather than just project executors, they secure stronger support. When I (Bucero) researched executive training in project sponsorship, I discovered most organizations do not prepare executives and make them aware of the need for working with project managers in order to contribute to project success.

IMPACT OF PM BELIEFS ON CUSTOMER RELATIONSHIP

A project manager's beliefs significantly impact customer relationships and service quality. Here's how:

1. Belief in Customer-Centricity
    - If a project manager believes that customer satisfaction is a top priority, they will actively engage with the customer, ensuring their needs are understood and met.
    - They focus on delivering value, leading to stronger relationships and long-term partnerships.
2. Belief in Collaboration and Teamwork
    - A project manager who believes in teamwork fosters better internal collaboration, ensuring smoother communication with the customer.
    - A cohesive team delivers better service, which enhances the customer experience.

3. Belief in Transparency and Trust
   - If the project manager values honesty and transparency, they will proactively update customers about project progress, risks, and challenges.
   - This builds trust and reduces conflicts, improving customer relationships.
4. Belief in Problem-Solving and a Positive Attitude
   - A project manager who believes in a solutions-oriented mindset will address customer concerns effectively.
   - Positivity and resilience in challenging situations reassure customers, strengthening their confidence in the service.
5. Belief in Continuous Improvement
   - A project manager who believes in learning and growth will seek feedback from customers and make necessary improvements.
   - This leads to enhanced service quality and long-term customer satisfaction.

In summary, a project manager's mindset shapes customer interactions, service quality, and the overall success of the relationship.

Would you like to explore how to develop these beliefs within a project team?

Tool

"Feedback on My Project Management Beliefs & Leadership Style"

Section 1: About You (Optional)

(These questions are optional, but they help contextualize the feedback.)

1. **What is your current role in the team or organization?**
   - Project Team Member
   - Project Sponsor
   - PMO Member
   - Stakeholder
   - Other (please specify)
2. **How long have you worked with me on projects?**
   - Less than 3 months
   - 3–6 months
   - 6–12 months
   - More than 1 year

Section 2: Project Management Beliefs

Please rate the following statements based on your experience working with me. (Scale: Strongly Disagree—Disagree—Neutral—Agree—Strongly Agree)

3. The project manager's positive attitude helped the team stay motivated.
4. I feel that people and relationships are prioritized in our projects.
5. The project manager adapts their leadership style based on team needs.
6. Recognition and feedback are given regularly and help me grow.
7. The belief in teamwork and collaboration is clearly reflected in project execution.
8. I feel comfortable approaching the project manager when I have a problem or need help.

Section 3: Impact and Areas for Growth

9. **What aspects of the project manager's approach have positively impacted you or the team?**

(Open-ended)

10. **Are there any project management beliefs or behaviors you feel could be improved?**

(Open-ended)

11. **How would you describe the overall working climate on the projects we've done together?**

(Open-ended)

12. **What's one piece of advice or encouragement you'd like to share with the project manager?**

(Open-ended)

Final Section (Optional)

13. Would you be open to a short follow-up conversation to clarify or expand on your feedback?
    - Yes (please leave your email)
    - No

HOW TO CREATE YOUR GOOGLE FORM

1. Go to https://forms.google.com
2. Click the **blank form** or **"+" button** to start a new form.
3. Use the following as your **form title**:

Feedback on My Project Management Beliefs and Leadership Style

4. Then copy and paste the questions and sections below into your form:

🗒 Form Questions and Layout

Section 1: About You (Optional)
Question 1: What is your current role in the team or organization?

Multiple Choice

- Project Team Member
- Project Sponsor
- PMO Member
- Stakeholder
- Other (please specify)

Question 2: How long have you worked with me on projects?

Multiple Choice

- Less than 3 months
- 3–6 months
- 6–12 months
- More than 1 year

Section 2: Project Management Beliefs *(Use "Linear scale" 1 to 5 for each)*
Instructions: *Rate the following statements (1 = Strongly Disagree, 5 = Strongly Agree)*

Question 3: The project manager's positive attitude helped the team stay motivated.

Question 4: I feel that people and relationships are prioritized in our projects.

Question 5: The project manager adapts their leadership style based on team needs.

Question 6: Recognition and feedback are given regularly and help me grow.

Question 7: The belief in teamwork and collaboration is clearly reflected in project execution.

Question 8: I feel comfortable approaching the project manager when I have a problem or need help.

Section 3: Impact and Areas for Growth
Use "Paragraph" for these open-ended questions.

Question 9: What aspects of the project manager's approach have positively impacted you or the team?

Question 10: Are there any project management beliefs or behaviors you feel could be improved?

Question 11: How would you describe the overall working climate on the projects we've done together?

Question 12: What's one piece of advice or encouragement you'd like to share with the project manager?

Final Section (Optional)
Question 13: Would you be open to a short follow-up conversation to clarify or expand on your feedback?

Multiple Choice

- Yes (please leave your email)
- No

Question 14 (if Yes): Please leave your email address:
Short Answer
Once the form is created, click **"Send"** and choose the **link icon** to share it with your team.

7

The Project Manager Mindset

There is no shortage of remarkable ideas, what's missing is the will to execute them.
Seth Godin

MINDFULNESS

The Oxford Dictionary defines mindfulness as the condition or attribute of being conscious of something. As a therapeutic method, it can also be defined as a mental state attained by quietly noticing and accepting one's thoughts, feelings, and physical sensations while concentrating one's awareness on the present. Being mindful. This is a simple term. It implies that the mind is completely focused on what is taking place, what you are doing, and the environment you are traversing. Aside from the fact that we frequently stray from the topic at hand, it can seem insignificant. Our minds take over, we get disconnected from our bodies, and before we know it, we're consumed by compulsive thoughts about the past or worrying about the future. And that makes us anxious. Mindfulness is the basic human ability to be fully present, aware of where we are and what we're doing, and not overly reactive or overwhelmed by what's going on around us. It's important that we don't define mindfulness too narrowly or confuse different meanings with each other. See Figure 7.1.

Unquestionably, mindfulness is a trendy term today, becoming as well-known as yoga. However, a thorough examination of the phrase "mindfulness" finds no one universal definition; rather, it is a complicated term

DOI: 10.1201/9781003730736-7

FIGURE 7.1
Mindfulness.
Source: Rawpixel.com/Adobe Stock.

that refers to several different but related concepts. When definitions are considered as interchangeable, this multiplicity leads to misunderstandings. Conflating several concepts of mindfulness also results in a lot of potential benefits being lost when it comes to critiquing. We propose some things we may be talking about when we talk about mindfulness:

1. A state of awareness with qualities: such as intentional, present, friendly, open, caring, and curious.
2. As opposed to the wider colloquial usage of neutral awareness of a concept or principle. (That is, the quality or state of being conscious or aware of something e.g. "their mindfulness of the mainstream cinematic tradition.")
3. A human trait or capacity: The tendency or ability to be in a mindful state as defined above. This trait is present to a greater or lesser degree in all of us—the basic human ability to be awake and aware of what we're doing, when we're doing it. This ability can be cultivated through practice.
4. Practice an intentional and regular act which familiarizes the participant with the qualities of mindfulness and develops their capacity to be mindful. This might be formal meditation, mindful pauses, or informally remembering to be mindful during the day.

5. A 'toolbox' of several practices: the word mindfulness is sometimes used to refer to a range of different practices including attention training, cultivation of qualities of heart, and even forms of visualization. Inquiry practices examining specific mechanisms of mind also yield ways of seeing the world that ease patterns of distress in the long term.
6. A training course or intervention that intentionally cultivates the trait (and state) of mindfulness. Training often includes a considerable amount of broader psycho-education and may accompany and/or catalyze cultivation of other capacities such as gratitude or compassion.
7. A quality of behavior often associated with being considerate; people sometimes joke "oops, that wasn't very mindful of me!" when what they mean is that they've been accidentally thoughtless or hurtful.
8. An umbrella term for an aspect of the dharma: the word dharma is used in Eastern contemplative traditions to refer to the 'ways things are' and any teachings that describe this. It involves an understanding of the mechanisms of psychological distress and the wisdom that leads to freedom from patterns of suffering.
9. A social movement: Mindfulness (with a capital 'M') as a cultural phenomenon is an example of a generative social field (a co-created climate or environment that fosters positive collective outcomes.)
10. A research outcome: building the evidence base for mindfulness entails the counterintuitive task of making it measurable. Scientists "measure" mindfulness using self-report questionnaires, brain scans, and other methodologies—but can this complex factor be quantified so simply? Mindfulness as a research outcome may tell us something about mindfulness as a state, capacity or practice, but shouldn't be confused with any of them: the map is not the territory.

EMBRACING AN OPEN MINDSET

We believe that an open mindset in project management means staying adaptable, receptive to new ideas, and willing to challenge your own assumptions. It enables you to navigate uncertainty, build stronger teams, and drive innovation while maintaining a people-focused approach to project success.

Projects are not delivered by methodologies. They are delivered by people. In the book *Today Is a Great Day*, I (Bucero) remark that attitude is a choice we make every morning. In *The Complete Project Manager* book, our message is equally clear: successful project managers combine technical competence with human leadership. An open mindset connects both worlds.

The Open Mindset Starts with Attitude: In *Today Is a Great Day* (Bucero, 2024), I emphasize that "today is a great day not because everything will go well, but because we choose how we face it." This belief is foundational for project managers operating in uncertainty. An open mindset allows project managers to accept reality without resistance and respond with intention rather than frustration. Optimism, in this context, is not about ignoring problems. It is about believing that every situation can be managed better with the right attitude. *The Complete Project Manager* book (Englund & Bucero, 2019) describes the project manager as someone who balances process, people, and purpose. An open mindset is what enables that balance. Rather than defending plans at all costs, the complete project manager remains open to:

- New stakeholder expectations
- Team insights that challenge assumptions
- Lessons learned from mistakes

We believe projects succeed when project managers focus less on control and more on influence.

Openness Builds Trust and Influence: Influence is earned, not imposed. An open mindset allows project managers to truly listen, especially when feedback is uncomfortable. Trust grows when people feel respected and heard. Open-minded project managers create psychological safety, enabling early risk identification and better decision-making. People do not commit to plans; they commit to leaders.

Practicing Openness Every Day: We encourage reflection and gratitude as daily habits. These practices reinforce openness by helping project managers:

- Reflect on what they learned each day.
- Focus on progress instead of setbacks!
- Recognize the contributions of others.

Small positive actions, repeated daily, create extraordinary results.

A Mindset That Defines Project Success: An open mindset transforms the project manager from a task controller into a people leader. It embodies the spirit of both books: positive attitude, human connection, and continuous learning. When project managers choose openness, they do more than deliver projects. They inspire teams, they influence organizations, and they prove—day after day—that today is a great day and tomorrow will be even better.

Manuel Ancizu, PMP shared with us his personal experience when managing projects:

> *There are two sentences I always keep in mind:*
>
> *1) "People are at the center of every project."*
> *2) "The only thing leaders truly have is their people and their team."*
>
> *Decisions must be made, agreements reached, and priorities aligned. Technology can certainly enhance efficiency and collaboration, but it is people who drive the project. They are the ones who go the extra mile to support a colleague, who push to meet a milestone, who propose the idea that changes everything, who motivate others, and who negotiate agreements that build trust and credibility. None of this happens automatically; it is people, with their willpower, determination, and desire to move the project forward, who make it possible. It is people who choose to involve and inspire others toward a shared objective that can positively impact the environment around them.*
>
> *A true PM mindset helps teams act as one, regardless of where they are located. It also recognizes that belief is the engine of progress: belief in the team's potential, in the capacity to overcome obstacles, and in the value of working together toward something meaningful. And, according to my experience, when people feel trusted, empowered, and connected, they can unleash their full potential and enable projects to create long-lasting impact.*

We found some key traits of an open-minded project manager, Figure 7.2, based on our practical experiences.

1. *Willingness to Learn and Unlearn*
    - Accepts that past experiences don't always define the best future approach.
    - Seeks new knowledge, industry trends, and alternative project methodologies.

FIGURE 7.2
Image of an open-minded project manager.

- Learns from team members, stakeholders, and even unexpected failures.

2. *Adaptability and Flexibility*
    - Adjusts strategies when project dynamics shift instead of resisting change.
    - Encourages creative problem-solving rather than sticking to rigid processes.
    - Embraces agility, using feedback to refine project execution.
3. *Active Listening and Open Communication*
    - Listens to understand, not just to respond.
    - Encourages diverse perspectives and constructive debate.
    - Avoids dismissing ideas just because they challenge conventional thinking.
4. *Growth-Oriented Decision Making*
    - Focuses on long-term improvement rather than short-term perfection.
    - Accepts that mistakes are learning opportunities, not failures.
    - Involves the team in decision-making, fostering ownership and engagement.
5. *Embracing Diversity of Thought*
    - Values different backgrounds, experiences, and perspectives.
    - Recognizes that innovation often comes from unexpected places.
    - Encourages cross-functional collaboration and knowledge sharing.

6. *Resilience and Positivity in Uncertainty*
    - Approaches challenges with curiosity rather than fear.
    - Maintains a constructive outlook even when plans don't go as expected.
    - Helps the team stay motivated by reframing obstacles as opportunities.

HOW TO DEVELOP AN OPEN MINDSET IN PROJECT MANAGEMENT

I (Bucero) follow my 3Ps (Passion, Persistence, and Patience) when managing projects in organizations, acting like a "mosquito" who tries to bite you smoothly but in a positive way. I (Englund) believe in changing thinking to help create organic environments for successful projects, in an environment where people can do their best work.

We suggest a set of actions that can help you develop an open mindset:

1. Ask More Questions: Replace assumptions with curiosity. Seek input from various sources before forming conclusions.
2. Challenge Your Own Beliefs: Regularly reflect on whether your current approaches are still the best ones.
3. Encourage Team Experimentation: Allow room for trial and error to drive innovation.
4. Stay Humble: Recognize that no one knows everything, including yourself. Learning never stops.
5. Give Constructive Feedback and Be Open to Receiving It: Growth happens when feedback is embraced rather than resisted.
6. Create a Safe Space for Ideas: Make sure team members feel comfortable proposing unconventional solutions.
7. Practice Active Adaptation: When new information emerges, be willing to shift direction if it benefits the project.

By embracing an open project manager mindset, you create a dynamic and inclusive project environment where continuous improvement, collaboration, and creativity thrive—leading to better results and a more engaged team.

CASE STUDY

Developing an Open Mind set in a Complex IT Transformation Project

Project Context: A multinational organization launched a 12-month IT transformation project to standardize systems across three regions. The project involved multiple vendors, culturally diverse teams, and an executive sponsor with high expectations and limited availability. Early project reviews showed increasing tension, missed milestones, and defensive behaviors within the team. The project manager, with strong technical credentials, realized that the project was not failing due to lack of knowledge—but due to lack of openness.

The Challenge

- Team members were reluctant to speak openly in meetings.
- Stakeholders defended their local interests instead of project goals.
- Risks were identified late, often when it was too late to act.
- The project manager felt pressure to "have all the answers."

The environment was driven by control, not trust.

The Mindset Shift
Inspired by the philosophy of *Today Is a Great Day*, the project manager made a conscious decision: change attitude before changing the plan. Instead of starting meetings with status updates, meetings began with one simple question: "What are we learning this week?"

This question signaled curiosity instead of judgment.

Actions Taken to Develop an Open Mindset:

1. Redefining Leadership Behavior
   The project manager openly admitted uncertainty when it existed and asked for input from the team. This reflected a key idea from *The Complete Project Manager*: leadership is about influence, not authority.

2. Creating Safe Conversations
   Short weekly reflection sessions were introduced, where team members shared:

   - One concern
   - One idea for improvement
   - One positive outcome from the week

   No actions were assigned during these sessions—only listening.

3. Partnering with the Project Sponsor
   Using principles about excellence in project sponsorship, the project manager engaged the sponsor differently: fewer reports, more dialogue. The sponsor was invited to listen to the team once a month without interrupting or evaluating.

Results Achieved
Within three months:

- Team engagement increased noticeably
- Risks were identified earlier and managed proactively.
- Cross-regional collaboration improved.
- The project recovered lost time without increasing budget.

More importantly, the team reported feeling trusted, heard, and respected.

Key Learning
The project did not improve because the plan was perfect. It improved because the mindset changed.

As reflected in *Today Is a Great Day*, a positive and open attitude creates the conditions for better decisions. And as emphasized in *The Complete Project Manager*, completeness comes from integrating technical discipline with human leadership.

Takeaway for Project Managers
An open mindset is not a personality trait—it is a daily leadership choice. When project managers model openness, teams follow. When teams open, projects succeed. The real transformation was not the IT system. It was the way people worked together.

THE MINDFUL PROJECT MANAGER

"The Mindful Project Manager" refers to a project management approach grounded in mindfulness principles, being fully present, self-aware, emotionally intelligent, and focused—while leading and managing teams and tasks. This concept emphasizes the *how* of project management as much as the *what* and *when*.

Key Principles of a Mindful Project Manager

1. **Presence and Focus**
   - Fully engages in the current task or conversation without distraction.
   - Avoids multitasking, helping reduce errors and improve quality.
2. **Self-Awareness**
   - Understands personal triggers, biases, and stress signals.
   - Recognizes when emotions may impact decision-making.
3. **Empathy and Active Listening**
   - Listens deeply to stakeholders and team members.
   - Build trust and collaboration by valuing diverse perspectives.
4. **Intentional Communication**
   - Chooses words and timing carefully to support clarity and reduce conflict.
   - Practices transparency and honesty with tact.
5. **Resilience and Stress Management**
   - Uses mindfulness techniques (like breath work or brief meditation) to manage stress.
   - Encourages a calm, steady leadership style in times of pressure.
6. **Ethical Leadership**
   - Makes decisions that align with values and long-term impact, not just deadlines and budgets.
   - Promotes sustainability and well-being across the project lifecycle.

Benefits in a Project Context

- There are fewer miscommunications and conflicts.
- Increased team engagement and morale.
- Better handling of uncertainty and change.
- More thoughtful decision-making and prioritization.

Practices for Mindful Project Management

- Start meetings with a brief check-in or moment of silence.
- Use reflective questions in retrospectives: *What did we learn? How did we feel?*
- Schedule "focus blocks" and protect them from interruptions.
- Maintain a personal mindfulness practice (meditation, journaling, etc.).

Aspirational Versus Delusional Thinking

We observe some people display a delusional mindset. Their key motivator is power, they want control, think they know everything and what needs to be done. Bullying others often occurs which makes others unwilling to confront them. They seem to hold false beliefs or opinions, even in the face of contradictory evidence. Active listening is not a regular occurrence.

In contrast, an aspirational mindset means having a strong desire for success or achievement. These people direct their thoughts toward a desired future with the belief that they can make it happen. Inspiring others is a regular byproduct of their behaviors. We gravitate toward these people and want to be around them. Through regular introspection, assess yourself whether either of these two mindsets guides your actions.

Here are comments from Eva Pareja Gómez, PMP® that illustrate mindset concepts in action:

> *My area of activity is asset management services (facility services). And not on the operations side, but on the sale of outsourced technical equipment maintenance (outsourcing) for properties such as hospitals, retail stores, office buildings, educational centers, etc. In the services area, each sales opportunity is a project with a beginning, an end, and specific milestones, where the sponsor is my own company, the solution is the deliverable, and the main stakeholder will be the client alone. Even if my project is outstanding, my solution would be accepted if chosen among the proposals (deliverables) of several competitors. Getting my proposal to become a finalist usually depends on strong teamwork. Still, my project can only succeed if the main stakeholder is satisfied and accepts it, thereby becoming my client.*
>
> *Focused on this point: perceived value depends solely on whether the project leader (the sales manager in this case) believes in the project, believes in the efforts of their team, believes in the organization's ability to carry it out, and shows it with passion at every stage. This enthusiasm is especially valuable in consultative selling, starting with gathering requirements and thoroughly*

understanding the potential client's underlying needs; continuing with working with the project technical team to convey an inspiring vision of value delivery; and finishing with a persuasive presentation aligned with the client's goals and internal expectations. Although most of the projects I get involved in are related to tenders (with rigorous selection criteria), this past year I have been fortunate to come across several projects where I have been able to apply this consultative selling methodology.

After several months of work and despite providing the least economically advantageous solution, I have earned the client's trust from three of them. All these success stories share a mindset of delivering value and enthusiastic communication.

Project Management Mindset Tools

There is a set of very practical tools that guide behavior, decisions, and relationships. Below is a clear, usable description of Project Management Mindset Tools, aligned our beliefs that people deliver projects, not processes.

1. Positive Framing Tool
   Purpose: Shift focuses on problems to possibilities.
   - Ask: "*What is still working?*" before "*What is broken?*"
   - Reframe risks as learning opportunities.
   - Replace "This will fail" with "What must be true for this to succeed?"
2. Stakeholder Empathy Map
   Purpose: Lead through understanding, not authority.
   - What does the stakeholder fear, need, and value?
   - What pressures are they under outside the project?
   - How do they define "success"?
3. Benefit-of-the-Doubt Tool
   Purpose: Build trust before conflict arises.
   - Assume positive intent when behavior is unclear.
   - Replace blame with curiosity: "*Help me understand.*"
   - Protect psychological safety within the team.
4. Gratitude and Recognition Practice
   Purpose: Strengthen motivation and engagement.
   - End meetings by acknowledging one contribution.
   - Encourage daily reflection on two positive actions.
   - Celebrate progress, not just results.

5. Open-Mind Questioning Tool
   Purpose: Avoid blind solutions.
   - Use questions like:
     - *"What am I not seeing?"*
     - *"Who should I listen to now?"*
     - *"What assumption might be wrong?"*

📌 Open minds reduce costly late-stage corrections.

6. Influence Before Authority Tool
   Purpose: Lead without forcing compliance.
   - Build credibility through consistency.
   - Communicate purpose, not just tasks.
   - Align decisions to shared goals.
7. Reflection Loop (Learn—Adjust—Move Forward)
   Purpose: Turn experience into wisdom.
   - Short, frequent retrospectives.
   - Focus on *learning*, not *fault.*
   - Ask: *"What will we do differently tomorrow?"*
8. Self-Leadership Check-In
   Purpose: Lead yourself before leading others.
   - How is my energy today?
   - What emotion am I transmitting to the team?
   - Am I reacting or responding?

Final Thought

Methodologies manage tasks. Mindset tools manage people, energy, trust, and decisions.

And as we often say: "Today is a great day — and because I choose the right mindset, tomorrow will be better."

Template

We propose a template or guide on how to integrate mindfulness into a real project plan. Mindfulness can help improve team focus, reduce stress, and enhance collaboration—key elements in successful project execution. The next section provides a practical **template and guide** to do that.

Tool

Integrating Mindfulness into a Project Plan

Objective: To enhance team well-being, focus, and performance through mindful practices.

1. Project Mindfulness Integration Plan Overview

Element	Description
Project Name	[Insert Project Name]
Project Manager	[Insert Name]
Mindfulness Champion	[Appoint a team member or external coach]
Start Date	[Insert]
End Date	[Insert]
Purpose	Foster clarity, reduce stress, and boost engagement across project phases.

2. Integration Points in Project Lifecycle

Phase	Mindfulness Activity
Initiation	Kick-off meeting with a 5-minute grounding exercise. Team sets shared intentions.
Planning	Weekly 10-minute breathing/centering sessions. Option: Mindfulness training workshop.
Execution	Mid-day pause (5–10 minutes) twice per week. "Mindful Mondays" start with silence.
Monitoring and controlling	Weekly check-ins include "How are you feeling?" reflection question.
Closure	Reflective review: what did we learn about ourselves and each other? Gratitude circle.

3. Mindfulness Practices to Use

- **Breath Awareness**: Focused breathing before stressful tasks or meetings.
- **Body Scan**: End-of-day body relaxation exercise (can be guided audio).
- **Silent Check-In**: Two minutes of silence to start meetings.
- **Mindful Communication**: Practice active listening, avoid interruptions.
- **Gratitude Sharing**: Weekly or daily positive sharing in team chats.

4. Roles and Responsibilities

Role	Responsibility
PM	Models mindful behavior; adds practices to calendar.
Team Members	Participate voluntarily; give feedback on usefulness.
Mindfulness Champion	Leads sessions; shares resources or guided practices.

5. Tools and Resources

- **Apps:** Headspace, Calm, Insight Timer
- **Book:** *Search Inside Yourself* by Chade-Meng Tan
- **Formats:** Guided audio, silent practice, journaling prompts

6. Success Measures

Metric	Target/Goal
Team Stress Levels	Monthly pulse surveys
Participation Rate	>70% team engagement in at least 1 weekly mindfulness activity
Feedback on Value	Qualitative survey post-project
Retention/Focus Improvement	Observations by project leads or peer feedback

Mindset as Your Superpower!

As seen in Figure 7.3, mindset is the power base of a pyramid representing progression of leadership skills. Work from bottom up. Many leaders believe

FIGURE 7.3

they are facing an accountability problem, when the real issue is a lack of self-accountability. Teams often fall into a "powerless" mindset shaped by blame, excuses, and the belief that outcomes are outside their control.

Since mindset is a set of assumptions, methods, or notions held by individuals and often rise out of a person's world view or philosophy of life, it's important to clarify your way of thinking and opinions, which influences mental inclination or tendencies. Focus first on developing an open mindset—this is your superpower. Then determine what skills are needed to support that mindset and only then gather tools to increase efficiency.

CONCLUSION

Remember that mindfulness is a basic human ability to be fully present, aware of where you are and what you're doing. Do not be overly reactive or overwhelmed by what's going on around you. Notice and accept your thoughts, feelings, and physical sensations while concentrating your awareness on the present. Basically, that means paying attention, not blind to the world around you. When mindfulness and emotional intelligence work together, more is possible in a quest toward excellence.

8

Summary and Conclusions

If you believe in yourself and have dedication and pride – and never quit, you'll be a winner. The price of victory is high but so are the rewards.

Bear Bryant

BELIEVE IN PEOPLE FIRST

We consider believing in your people is an investment. We suggest believing in your people, and you will get a surprising return of that investment. We summarize our learnings as follows:

- *Trust that your team members are capable, committed, and willing to grow:* You, as a project manager need to trust your people and give to them the benefit of the doubt. In my experience (Bucero) I have obtained more from those people when I opened my mind to understand them, listen to their stories, and learn from their reactions.
- *Empower them by delegating responsibilities and recognizing their efforts*: Always delegate with accountability. When people are committed to their assigned work and get it finished, they need to be recognized.
- *Your belief in their abilities will boost their confidence and performance*: Believe everyone in your team is special. If you deal with them as especial ones, they will surprise with their achievements and level of performance.

 DOI: 10.1201/9781003730736-8

BELIEVE IN THE PURPOSE OF THE PROJECT

Project purpose is the fundamental reason why a project exists. It describes what the project aims to achieve and why it matters—both to the organization and to the stakeholders involved. It connects the project to a broader strategic goal or need and serves as a guiding star throughout the project life cycle. Here are characteristics of a project purpose:

- **Strategic Alignment**: It ensures the project supports the organization's mission, vision, or business strategy.
- **Problem-Solving or Opportunity-Driven**: The purpose usually addresses a problem to be solved or an opportunity to be seized.
- **Motivational Role**: It helps engage and inspire team members and stakeholders by clarifying the value of the work.
- **Decision-Making Anchor**: It helps prioritize tasks, resolve conflicts, and evaluate project changes.

For example, if the project is to implement a new customer relationship management (CRM) system, the **purpose** might be:

"To improve customer satisfaction and increase sales effectiveness by providing sales and service teams with a unified, efficient, and unique platform for managing customer interactions."

- Ensure you understand and communicate the project's purpose and value. If you don't believe in the project's impact, it will be hard to motivate others. Align the team with a meaningful vision to maintain long-term motivation.

BELIEVE IN ADAPTABILITY AND RESILIENCE

- Accept that changes, setbacks, and uncertainty are natural parts of any project.
- Instead of resisting challenges, embrace them as opportunities to improve.
- Lead by example—stay calm, adaptable, and solution-focused.

BELIEVE IN CONTINUOUS IMPROVEMENT

- See every project as a learning opportunity.
- Encourage feedback and self-reflection to refine processes and skills.
- Recognize that mistakes are part of growth, not a sign of failure.

BELIEVE IN TRANSPARENCY AND HONEST COMMUNICATION

- Trust that open communication builds stronger teams and stakeholder relationships.
- Be honest about risks, expectations, and challenges while remaining optimistic.
- Foster an environment where people feel safe to voice concerns and ideas.

BELIEVE IN A POSITIVE ATTITUDE

- Positivity doesn't mean ignoring problems—it means focusing on solutions.
- Encourage gratitude and reflection (as you already do with your team).
- Celebrate small wins to maintain motivation and momentum.

BELIEVE IN THE VALUE OF STAKEHOLDER ENGAGEMENT

- Strong project sponsorship and executive support can make or break a project.
- Invest time in building relationships with sponsors and key stakeholders.
- Advocate for the project's importance in a way that gains their trust and commitment.

BELIEVE IN BALANCE AND WELL-BEING

- A healthy mindset requires maintaining personal well-being and avoiding burnout.
- Encourage work-life balance within the team to sustain long-term performance.
- Recognize when stress is affecting decision-making and take steps to reset.
- Create an environment of psychological safety.

Since you already focus on motivating people, recognizing their contributions, and maintaining a positive outlook, these beliefs likely align with your leadership style.

We recommend adopting a **circular thinking approach** to create a foundational strategic shift: stop thinking in terms of monetary or output results; instead start thinking in terms of the value that meetings, events, and interactions can create. Use resources deployed in the best possible way. Embrace and propagate this shift.

Here is a suggested **Google Forms structure**. Copy and paste the following into Google Forms manually or we can provide shareable content in spreadsheet format to import if needed.

Tool

Project Manager Beliefs and Best Practices Survey
Form Description
This survey is designed to evaluate how well project managers and team members align with key beliefs that support successful project leadership. Honest responses will help identify strengths and areas for development.

Section 1: Belief Alignment (1 = Strongly Disagree, 5 = Strongly Agree)
(Use a **Linear Scale** question format from 1 to 5 with labels: "Strongly Disagree" and "Strongly Agree")

1. **Belief in People First**
   - I trust that my team members are capable and committed.
   - I delegate responsibilities and recognize team efforts regularly.
   - My belief in my team improves their performance.

2. **Belief in the Purpose of the Project**
   - I fully understand the purpose and value of our projects.
   - I communicate that purpose clearly to the team.
   - I feel motivated because I believe the project has meaningful impact.
3. **Belief in Adaptability and Resilience**
   - I see challenges as opportunities to improve.
   - I remain calm and focused when projects face setbacks.
   - I encourage the team to adapt to change positively.
4. **Belief in Continuous Improvement**
   - I view each project as a learning experience.
   - I actively encourage feedback and reflection.
   - Mistakes are treated as opportunities for growth, not failures.
5. **Belief in Transparency and Honest Communication**
   - I communicate risks and challenges openly.
   - I foster an environment where concerns can be voiced safely.
   - I practice optimism while staying honest with my team.
6. **Belief in a Positive Attitude**
   - I maintain a solution-focused mindset.
   - I encourage gratitude and reflection within the team.
   - I celebrate small wins to keep morale high.
7. **Belief in Stakeholder Engagement**
   - I regularly engage with project sponsors and key stakeholders.
   - I advocate for the project's importance to gain executive support.
   - I believe strong sponsorship is critical to project success.
8. **Belief in Balance and Well-Being**
   - I prioritize my own well-being to maintain leadership effectiveness.
   - I support work-life balance among team members.
   - I take proactive steps to manage stress and avoid burnout.

Section 2: Open-Ended Questions
(Use **Paragraph** response type)

9. Which belief do you consider your strongest area as a project manager? Why?
10. Which belief do you find most challenging to uphold in your current projects?
11. Share an example where applying one of these beliefs helped overcome a project obstacle?
12. What support or training would help reinforce these beliefs further?

A point of view was posted by Kayvon Farsad from Northeastern University, whose belief changed and started saving projects from going off track and preventing [negative] risks from becoming reality:

> For most of my career, I thought my value was developing plans, timelines, identifying deliverables, and handing over checklists. HOWEVER, I've discovered that it's something way way way deeper than that.
>
> I am referring to THE ABILITY TO SIT INSIDE AMBIGUITY WITHOUT PANIC.
>
> Yes, I learned it the hard way, through managing projects where the requirements weren't real yet, working across enterprise teams that didn't establish trust for cross-functional deployments, teaching while still creating new knowledge systems, and rebuilding my professional identity when things fell apart.
>
> What I loved about these moments, even when they were brutal, is that they FORCED ME TO STOP PRETENDING, and intentionally focus on building clarity incrementally. That could mean breaking vague goals into testable, tiny experiments, asking better questions instead of faster reactions, creating structure that calms people down (instead of creating more pressure in their already complex workflows), and making progress and actions performed visible when outcomes aren't yet apparent.
>
> That way, we can easily verify what's happening and validate what's working (and equally important, what's not working).
>
> I choose to work this way because projects fail when stakeholders avoid hard issues and actions, not because of a lack of digital tools, capital assets, or even human resources, like you and me. That's why it's more important than ever not to rush processes and realize it's not the skills on your resume that distinguish you. It's actually the roadmap you quietly created that people keep coming back to you for.

Another posted point of view comes from Zack Gould:

> There are two lines that hurt my soul.
>
> I hear them constantly.
>
> In sports and in business.
>
> "They don't want it enough."
>
> I hear it from coaches in youth athletics all the time.
>
> And it makes my skin crawl.

> We're talking about 8-year-olds.
> 10-year-olds.
> High school kids.
> That line doesn't diagnose the problem.
> It breaks trust between the coach and the team.
> Then there's the business version.
> "They don't care like we do."
> I hear it in leadership meetings.
> From owners.
> From managers talking about underperforming teams.
> It sounds different, but it's the same sentence.
> When you say someone "doesn't care" …
> You're describing the gap between what you expect and what you've actually taught.
> Own the gap.

This story reminds me (Englund) of a professional football coach who chastised instead of congratulated a defense player who knocked the ball away rather than intercepting a key pass play. The coach said, "You didn't want the ball enough to go after it." The same approach applies for key belief constructs—fully embrace a mindset to adopt, adapt, and apply beliefs that support personal and professional development. Know that this mindset is your superpower to achieve consistent and sustainable success!

We conclude with a few words from trusted colleagues:

> I built a successful career without a traditional path by believing in myself and leaning on strong mentors. That same mindset—belief, positivity, and connection—is what makes *Believe: The PM Mindset* so powerful. This book offers real-world insights every project manager can use to thrive.
>
> (Lisa DiTullio lisad@projectchalktalk.com)

Ricardo Vargas, Ph.D., PMI Fellow adds:

> *The PM Mindset is a powerful reminder that project success starts long before tools, methods, or frameworks—it begins with belief. This book goes straight to the core of what truly differentiates great project managers: mindset. Through real stories, practical reflections, and lessons from the field, it shows how believing in yourself, your profession, and your teams unlocks levels of performance that technique alone can never reach. This is not a theory. It is a lived project management, written with authenticity, optimism, and courage—a valuable read for anyone who wants to achieve more by first believing more.*

Vivki James writes in her blogs about Age, Choice, and the Life We Build:

What makes this possible is the accumulation of choices I've made at each hard turn in the road—choices to adapt, to learn, and to believe I deserved and was capable of something better.

That's what resilience has looked like in my life. Not grit for its own sake, and not fearlessness—but deciding, again and again, to act on the **belief** that there was another way forward.

The Stories That Keep Us Still: When people talk about why they don't make changes, I often hear variations of the same themes:

- It's too late to start over
- This is just how things are at this stage of life
- I can't afford the risk
- I don't have the energy I used to
- People will think I'm foolish

I understand these thoughts deeply. They sound practical. Responsible. Mature. But I've learned that many of them are less about truth and more about protection. Over time, our minds build filters—shaped by experience, disappointment, and well-meaning advice. Those filters can quietly turn fear into something that looks like comfort. We absorb phrases like "the grass isn't always greener" or "better the devil you know," and they settle in as unquestioned rules. Sometimes what feels like wisdom is simply familiarity doing its job.

The Myth of Being "Special": Another subtle trap sounds like this:

That might work for someone else, but it could never work for me.

I've believed that, too. It's tempting to think our circumstances are uniquely heavy, uniquely complex, uniquely limiting. And sometimes they are. But what we often don't see is how isolating that belief can become.

I've witnessed this clearly in recovery spaces. People arrive with very different stories, but the turning point is often the same: letting go of pure self-reliance and becoming willing to receive help.

When we're fully self-reliant, we only have access to the options we already know. When we allow support, new paths appear—paths shaped by people who have been where we are and learned how to make different choices.

Someone, somewhere, has stood at a crossroads similar to yours and found a way forward. Not because they were braver or luckier—but because they were willing to be honest, curious, and supported.

Resilience isn't about powering through or pretending things don't hurt. It's about deciding—again and again—that you deserve something better and being willing to act on that belief by seeking support and accepting help. Because if the only information you're relying on is what you already know, there's no room for something new to enter.

Your Turn: If there's a dream you quietly shelved along the way, you don't have to resurrect it all at once. You don't even have to know the full path forward.

You might start by asking:

- When did I decide this wasn't possible anymore?
- What support would help me take one small step toward it?
- If comfortable, share here so we can support each other.

Change doesn't usually arrive as a dramatic leap. More often, it shows up as a series of steady, imperfect choices—made at hard turns in the road.

That's what resilience has looked like in my life.

We thank Vicki … and all our readers, and we offer best wishes throughout your quest to believe in and receive all that life can offer.

Here is a summary mind map for this book, Figure 8.1:

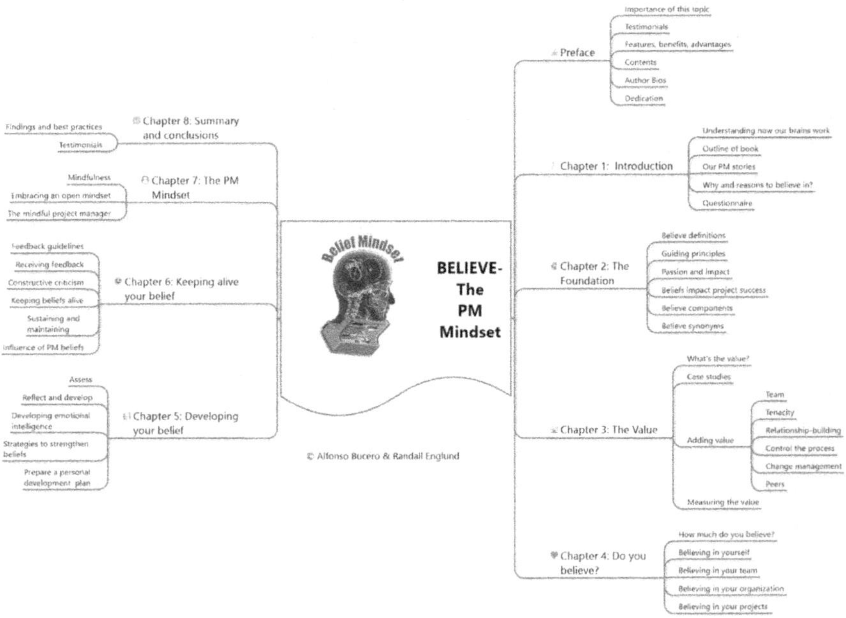

FIGURE 8.1
Believe mind map.

Appendix

Tools for the Project Management Believer

Inspired by the works of Randall L. Englund and Alfonso Bucero

INTRODUCTION: WHAT IS A PROJECT MANAGEMENT BELIEVER?

A project management believer is someone who truly trusts that

- Projects are vehicles for strategy execution
- People—not processes—drive success
- Leadership, attitude, and influence matter as much as plans and tools

Randall L. Englund and Alfonso Bucero consistently reinforce a simple truth:

Great project results come from committed sponsors, supportive environments, influential project managers, and engaged teams—working together with purpose. This Appendix presents mindset tools, leadership tools, and relationship tools that help reinforce belief in project management as a discipline and a profession.

Belief Tool #1: Positive Attitude as a Strategic Asset

(Bucero – *Today Is a Great Day*)

What It Is
A conscious choice to approach projects with optimism, realism, and energy—even under pressure.

Why It Matters

Belief begins internally. Project managers who project confidence and positivity influence:

- Team motivation
- Stakeholder trust
- Sponsor engagement

Practical Tool

Daily Reflection Practice

Ask yourself and your team:

- What went well today?
- Who helped us succeed?

Reinforces momentum and learning.

Today is a great day—because I choose to make it one.

Alfonso Bucero

Belief Tool #2: Sponsor Engagement Framework

(Englund and Bucero – *Project Sponsorship*)

What It Is

A structured approach to ensuring executives actively support projects—not passively approve them.

Why It Matters

Projects fail not because of poor scheduling—but because of weak sponsorship.

Practical Tool

Sponsor Commitment Checklist

- Clear ownership of business outcomes
- Visible executive support
- Timely decision-making
- Active participation at key milestones

This tool strengthens belief that projects are leadership-driven, not just manager-driven.

Belief Tool #3: Influence without Authority

(Bucero – *The Influential Project Manager*)

What It Is
The ability to lead people you do not formally manage.

Why It Matters
Project managers rarely have full authority—belief comes from mastering influence.

Practical Tool
Influence Map

- Identify key stakeholders
- Understand their interests, fears, and success criteria
- Adapt communication style intentionally

Believers don't wait for power—they create alignment.

Belief Tool #4: The Complete Project Manager Model

(Englund and Bucero – *The Complete Project Manager*)

What It Is
A balanced model integrating:

- Technical competence
- Leadership, influence, and personal skills
- Business awareness
- Personal integrity

Why It Matters
Belief grows when project managers stop seeing themselves as "task coordinators" and start acting as business leaders.

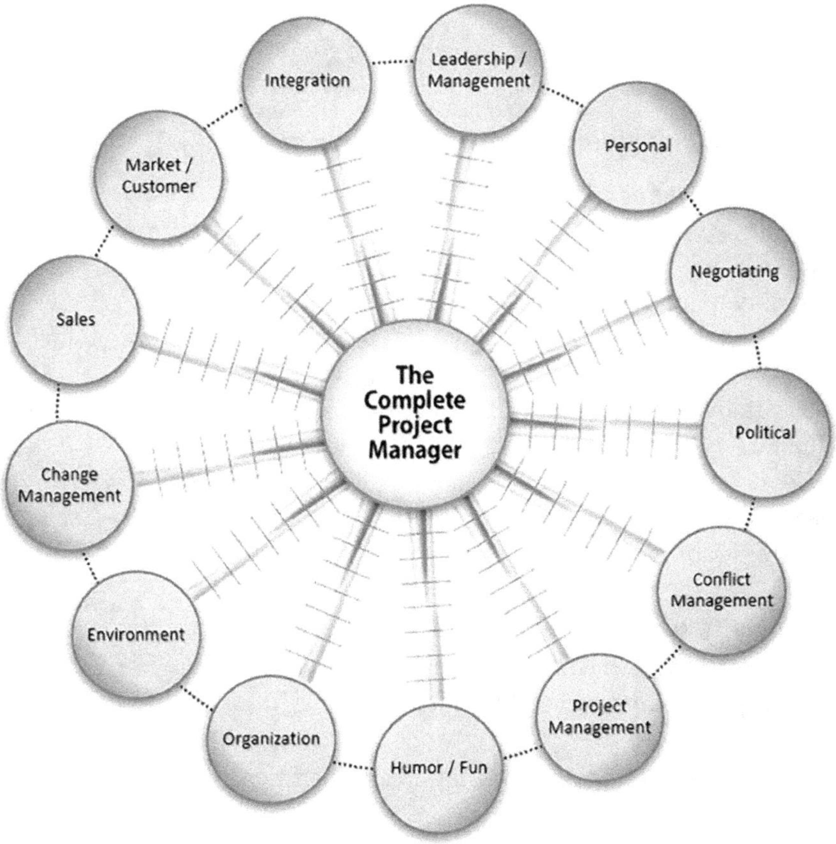

FIGURE A.1
Complete project manager molecule.

Practical Tool

Self-Assessment Wheel. See Figure A.1.
Evaluate yourself across:

- Leadership
- Communication
- Business acumen
- Emotional intelligence
- Delivery discipline

Gaps become development opportunities—not weaknesses.

Belief Tool #5: Passion, Persistence, and Patience (The 3Ps)

(Bucero – *Passion, Persistence, and Patience*)

What It Is
A mindset trilogy for long, complex, and politically challenging projects (also applies to most endeavors).

Why It Matters
Belief is tested over time.

Practical Tool
3P Reflection

- Passion: Why does this project matter?
- Persistence: What obstacles must I keep pushing through?
- Patience: What cannot be rushed?

Projects succeed not by speed alone—but by steady belief and resilience.

Belief Tool #6: Trust-Based Team Building

(Both authors—recurring theme)

What It Is
Creating psychological safety and mutual respect within the project team.

Why It Matters
Belief spreads socially. Teams believe in projects when they believe in each other.
Rotating Leadership and Recognition

- Rotate meeting facilitation
- Publicly recognize contributions
- Encourage peer appreciation

Trust fuels commitment.

Belief Tool #7: Projects as Strategy Execution

(Englund – *Creating an Environment for Successful Projects*)

What It Is

Positioning projects as the primary mechanism to deliver organizational strategy.

Why It Matters

Executives believe in project management when they see it as strategic, not operational, and embrace an organic environment.

Practical Tool

Strategy-to-Project Alignment Canvas

- Strategic objective
- Project contribution
- Success metrics
- Sponsor accountability

This reframes projects as value-creation engines. Figure A.2 depicts pieces of an environmental puzzle–all are important when creating a consistent, sustainable environment for successful projects.

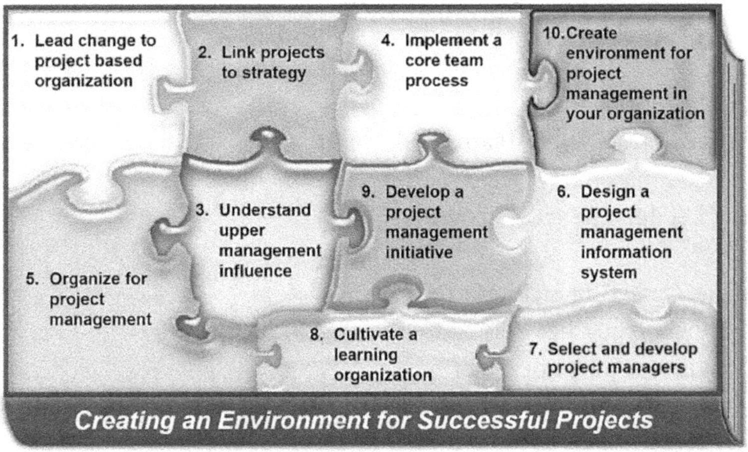

FIGURE A.2
Environment puzzle.

CONCLUSION: BECOMING—AND STAYING—A BELIEVER

Belief in project management is not blind faith.
It is built through:

- Leadership behaviors
- Sponsor commitment
- Positive attitude
- Influence and trust
- Continuous reflection

As Randall L. Englund and Alfonso Bucero demonstrate through their work:

When people believe in projects, projects deliver results.

And when project managers believe in themselves—today truly is a great day and tomorrow will be even better.

Authors' Books as Further Reference Material

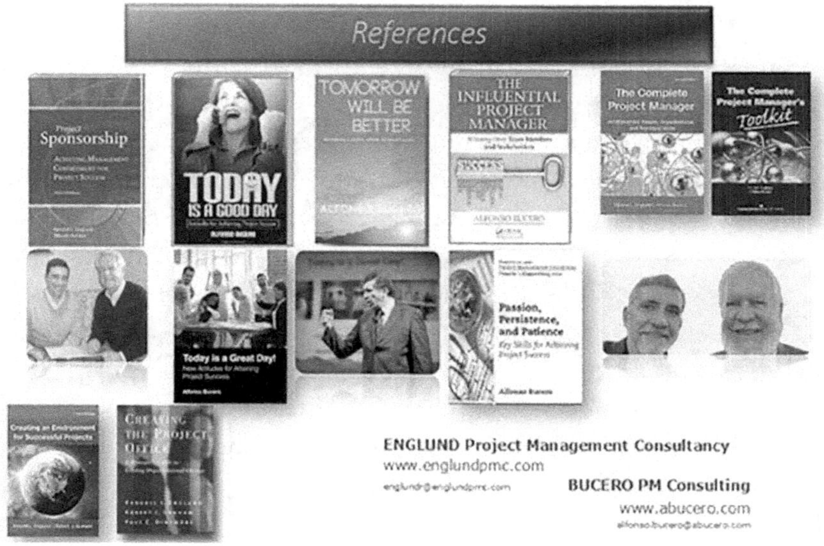

FIGURE A.3
Author reference books.

BELIEVE ESSAY ON PROJECT SPONSORSHIP

I'll believe it when I see it.
I'll see it when I believe it.

Do upper managers only come to believe in project sponsorship when they see it in action, perhaps from others in their organization or elsewhere (social proof)? Or do they believe in the vitality of project sponsorship and want to see it better implemented in their organization?

The latter belief is a good motivator for action.

Most executives we know seem to not believe in the need for project sponsorship, for several reasons. Some examples are as follows:

1. What is in it for me?
2. The person responsible for the project is the project manager.
3. As an executive I need to ask for the results; it's up to you to complete your project!
4. I am very busy, do not bother me.
5. You as project manager lead your team and need to get your project done.

As a project manager, I (Bucero) worked with executives in organizations explaining the project sponsor role and how the project sponsor can add value to the organization, to the project manager, and to the project. I clarified when, what, and how the project sponsor needs to participate in a project to contribute to project success, but many executives say *I'll believe it when I see it.*

Giving the benefit of the doubt needs to be, in our perspective, an obligation for the executive to perform the project sponsor role, as a leader. The lack of trust and dialogue between executives and project managers seems to be a constant situation in many organizations. The project sponsor assigns the project manager to a determined project, expedites the project charter and has a meeting with the project manager, then disappears till the first project milestone arrives. Playing the role of a "**ghost**" as a project sponsor is not desirable. Project sponsors and project managers need to talk periodically and be aligned because they need to help each other to the benefit of the project and the organization. To achieve a successful project, they need to work as a team. As much as they establish better dialogue, the better they will understand each other.

Project sponsors are strategically busy, and project managers are tactically busy. Why not establish a link between their tactic and strategic work? Until that happens, they are condemned to not work together to the detriment of the project and the organization. Many questions that come to the project manager's mind would be answered ASAP when the dialogue link between those roles works well.

The need from executives to see results to believe in is there. Project managers need to find arguments and justifications that may generate more and more awareness about the need for the project sponsor role.

On the other hand, project managers that believe in their projects, work closer and closer with their executives and make the effort to create awareness of the need for project sponsorship. Those people defend the thought: *I'll see it when I believe it.* We (Englund and Bucero) believe in our profession and believe that working together with executives as a team generates better project outcomes.

I (Bucero) conducted research about executive training in project sponsorship. In surveys submitted by 410 people worldwide, close to 60% of those people said that in their organizations their executives never received any project sponsorship training.

Executive training in project sponsorship is a must to have executives believing in the need for project sponsor involvement from beginning to end of their projects. There is still a lack of project sponsorship training in organizations worldwide, where project managers and executives can learn how to work together for project and organizational success. See Figure A.4, depicting steps to achieving excellence in project sponsorship.

How frequently are you talking to your executives and urging them that you need their involvement to achieve project success? Since some of them ignore it, be a kind, persistent "mosquito" reminding them they are critical for project success and the future survival of their organizations.

Believe in the power of commitment to make a difference. Use Figure A.5 as a regular mantra.

FIGURE A.4
Steps to project sponsorship excellence.

WHEN YOU COME TO BELIEVE,

YOU CAN ACHIEVE.

FIGURE A.5
Believe and achieve.

Checklists for these and other resources are available from https://englundpmc.com. Get more from BUCERO PM Consulting at https://abucero.com.

Index

Note: Page locators in **bold** and *italics* represents tables and figures, respectively.

For Product Safety Concerns and Information please contact our EU
representative GPSR@taylorandfrancis.com
Taylor & Francis Verlag GmbH, Kaufingerstraße 24, 80331 München, Germany

www.ingramcontent.com/pod-product-compliance
Lightning Source LLC
LaVergne TN
LVHW020940160926
848142LV00030B/290
* 9 7 8 1 0 4 1 2 2 1 2 1 0 *